An Introduction to TRANSFORM THEORY

This is Volume 42 in
PURE AND APPLIED MATHEMATICS

A Series of Monographs and Textbooks

Editors: Paul A. Smith and Samuel Eilenberg
A complete list of titles in this series appears at the end of this volume

An Introduction to TRANSFORM THEORY

D. V. WIDDER

Department of Mathematics
Harvard University
Cambridge, Massachusetts

1971

 ACADEMIC PRESS   New York and London

ACADEMIC PRESS, INC.
111 Fifth Avenue, New York, New York 10003

United Kingdom Edition published by
ACADEMIC PRESS, INC. (LONDON) LTD.
Berkeley Square House, London W1X 6BA

LIBRARY OF CONGRESS CATALOG CARD NUMBER: 79-154399

AMS (MOS) 1970 Subject Classifications: 44-01, 44-02, 44A05,
44A10, 44A35, 10H05, 10H15, 30A16.

PRINTED IN THE UNITED STATES OF AMERICA

Contents

3. The Zeta Function

4. The Prime Number Theorem

5. The Laplace Transform

9.   Inversion by Series

Bibliography   243

Preface

This book is essentially compiled from notes on lectures given by the author at Harvard University in a half-course on transform theory. It was attended chiefly by seniors and first-year graduate students, and only a basic knowledge of real and complex function theory was assumed. The book is designed to touch on a variety of the most fundamental aspects of the theory rather than to strive for encyclopedic coverage of any part. We hope that it will be useful to a student who is sampling various kinds of mathematics before settling on a direction for his own research.

The text begins with a rapid introduction of the use of Laplace integrals for solving differential equations. Although emphasis throughout is on the theoretical rather than on the applied side of the subject, any student of transform theory will wish to be cognizant of this most important application.

The basic properties of Laplace integrals can be conjectured by analogy from those of Dirichlet series. Consequently our theory begins with a chapter on such series. Since this "discrete" transform does not present some of the complications of the continuous, or integral, transform, it offers good introductory material. The most famous Dirichlet series is probably the one defining the zeta-function of Riemann. It is also the simplest in some ways since all the coefficients are unity. Yet it remains an enigma in that its zeros have not yet been completely located. Its tremendous influence on mathematics over the years almost makes its study obligatory for all mathematicians and certainly for students of analysis and number theory. Its basic properties, especially those needed later, are collected in Chapter 3.

Chapter 4 gives a proof of the prime number theorem, as one important application of Dirichlet series. To understand it the reader need have no previous knowledge of number theory. The material begins with Tchebychev's derivation of the order of magnitude of the nth prime although this is unnecessary for the main theorem. But this historical approach serves to give an introduction to the methods of number theory to familiarize the student with the number theoretical functions involved and to give him a better appreciation of the final result.

Although Dirichlet series form ideal introductory material, the student who wishes to immerse himself at maximum speed into the theory of integral transforms may omit Chapters 2–4, and proceed directly to the rest of the book. Chapter 5 sets forth the classic results about Laplace and Stieltjes transforms. The following chapter takes up the more recent inversions of these transforms, after first developing the Laplace asymptotic method. The latter is an indispensable tool for analysts and applied mathematicians.

In Chapter 7 a very rapid approach to the convolution transform is to be found. This basically subsumes earlier results and should serve to solidify the reader's understanding. The reason for the success of the earlier inversion formulas becomes apparent as they are recaptured in this more general setting.

Chapter 8 endeavors to introduce the reader to Tauberian theorems rapidly and simply. Two approaches are taken: one, via the general Tauberian theorem of N. Wiener [1933], the other through Karamata's specialized method. The former is for general kernels but is restricted to two-sided Tauberian conditions, the latter is for special kernels but permits the more general one-sided conditions. It is noteworthy that no use of Fourier analysis is made. This is avoided by our introduction of the uniqueness class U, to which the kernels here considered are already known to belong. The classic series theorems of Hardy and Littlewood are extracted as special cases.

We hope that the final chapter will prove intriguing to the reader, perhaps stimulating him to investigate more general results in the same direction. We present here amusing algorithms for the inversion, by series, of two special transforms. But the method is general, as the author has shown.

Exercises appear at the ends of chapters, some with answers. They are usually simple, intended to help the reader to test and to solidify his mastery of the text.

Theorems are generally stated in the same systematic and compact style used by the author in his "Advanced Calculus." The few logical symbols employed to accomplish this are for the most part self-explanatory, but a few are explained parenthetically when introduced for the first time. A separate index of symbols and notation can be found on pp. xiii and xiv.

Symbols and Notation

Page			
98	$\sigma_c{}'$		
98	$\sigma_c{}''$		
104	V	bounded variation	
107	V^*	normalized bounded variation	
108	B	bounded	
111	$*$	convolution	
122	$\{\rho, \gamma\}$	growth of an entire function	
136	$\downarrow$	nonincreasing	
140, 225	$L_{k,y}[\ ]$	inversion operator	
143	$M_{k,y}[\ ]$	inversion operator	
149	c.m.	completely monotonic sequence	
150	$M[\ ]$	moment operator	
150	$N[\ ]$	moment operator	
151	$F_{m,\,n}(x)$		
154	c.m.	completely monotonic function	
161	C^∞	continuous with all derivatives	
171	e^{aD}	translation operator	
172	$E(D)$	inversion operator	
173	E	entire (Laguerre–Pólya class)	
174	E_0	entire (Laguerre–Pólya subclass)	
175	M_p	center of gravity	
175	V_p	moment of inertia	
190	$\exp(tD^2)$		
191	$\exp(-tD^2)$	inversion operator	
193	(A)	summable Abel	
194	(C)	summable Cesàro	
199	U	uniqueness class	
204	SO	slowly oscillating	
209	SD	slowly decreasing	
211	C^n	continuous with first n derivatives	
225	$\theta[\ ]$	differential operator	

1 *Introduction*

1. Introduction

In this chapter we shall introduce the Laplace transform in the simplest possible setting with a view to showing, at the outset, a few of the possible applications. The chief emphasis of the book will be on the theoretical rather than on the applied side of the subject. But any student of transform theory will probably wish to be cognizant of the many possible applications. Without learning a vast technique he can at once appreciate the methods of solving linear differential equations with constant coefficients, for example. For this, no more complicated mechanism is needed than the process of integration by parts. A table of Laplace transforms is essential, and we shall begin by deriving a primitive one. A more extensive one is of course needed for the more complicated differential equations and for other applications, but many of these are now available in book form. Such tables are easily used once the method is understood.

In brief, the procedure is this. The Laplace transform is applied to both sides of the given differential equation. The result is an equation which can be solved algebraically. The solution of the algebraic equation is then the Laplace transform of the desired solution of the differential equation. The latter can then be found, at least in the simple cases here envisaged, by an inverse use of the table of transforms. The process is akin to the solution of an arithmetic problem by use of a table of logarithms. The first application of the table reduces one operation (such as multiplication) to a simpler one (addition). After the simpler problem is solved an inverse use of the tables yields the required solution.

We begin with a formal definition.

Definition 1. The Laplace transform of a function $\varphi(t)$ is the function

$$(1.1) \qquad f(s) = \int_0^\infty e^{-st}\varphi(t)\,dt.$$

Here $\varphi(t)$ is called the *determining* function, and $f(s)$ is the *generating* function.

Since the integral (1.1) is improper, a question of convergence arises. We shall see later, Chapter 5, that the integral always converges, if it converges at all, on a right half-line, (a, ∞), if s is real, or in a right half-plane if s is complex. Certain functions, like $\varphi = \exp t^2$, have no transforms f, since (1.1) may diverge for all s. Also there are certain functions, like $f = 1$ or $f = s$, that cannot be generating functions. For, it is easy to see from Eq. (1.1) that $f(+\infty) = 0$, for example. This property alone excludes a host of candidates from the rank of generating function.

For (1.1) to have meaning it is clearly sufficient for $\varphi(t)$ to belong to class L (Lebesgue-integrable) on $(0, R)$ for every $R > 0$ and for the improper integral to converge. However, for the purposes of the present chapter we shall assume only that $\varphi(t) \in C$ (is continuous) on $(0, \infty)$ and that (1.1) converges for some s. As a first trivial example we see that if $\varphi(t) = 1$, then $f(s) = 1/s$ and the transform converges for $s > 0$.

One further fact which we shall need at once, but which will not be proved until Chapter 5, is the uniqueness of the representation (1.1). That is, a generating function cannot be the transform of more than one continuous determining function. We state this result, in equivalent form, as a theorem.

Theorem 1.   1. $\varphi(t) \in C \qquad 0 < t < \infty;$

2. $f(s) = \int_0^\infty e^{-st}\varphi(t)\,dt \equiv 0 \qquad a < s < \infty, \quad \text{some } a$

$\Rightarrow \qquad\qquad \varphi(t) \equiv 0 \qquad 0 < t < \infty.$

2. A Brief Table of Transforms

The following brief table of transforms will be useful in the solution of the simple problems proposed in this chapter. It is a miniature of the vast tables now available. See, for example, A. Erdélyi [1954]. We regard s as a real variable in this chapter.

$$\int_0^\infty e^{-st}\,\varphi(t)\,dt = f(s)$$

$\varphi(t)$	$f(s)$	Conditions
1. t^{a-1}	$\Gamma(a)s^{-a}$	$a > 0, \quad s > 0$
2. e^{at}	$(s-a)^{-1}$	$s > a$
3. $\sin at$	$\dfrac{a}{s^2 + a^2}$	$s > 0$
4. $\cos at$	$\dfrac{s}{s^2 + a^2}$	$s > 0$
5. $\sinh at$	$\dfrac{a}{s^2 - a^2}$	$s > |a|$
6. $\cosh at$	$\dfrac{s}{s^2 - a^2}$	$s > |a|$
7. $t \sin at$	$\dfrac{2as}{(s^2 + a^2)^2}$	$s > 0$
8. $\sin at - at \cos at$	$\dfrac{2a^3}{(s^2 + a^2)^2}$	$s > 0$

These transforms are all derived by the familiar methods of advanced calculus. Let us recall the procedures. A change of variable gives

$$\int_0^\infty e^{-st}t^{a-1}\,dt = \frac{1}{s^a}\int_0^\infty e^{-t}t^{a-1}\,dt, \qquad s > 0,\ a > 0,$$

and the integral on the right is the gamma function, defined for $a > 0$. In particular

$$\int_0^\infty e^{-st}\,dt = \frac{1}{s}, \qquad s > 0,$$

as noted in §1. A translation of s through a gives formula 2. Note that we may always replace s by $s - a$ in a generating function if we multiply the corresponding determining function by e^{at}. This gives an enlargement of our table. If we note that formula 2 is still valid for complex numbers a, then all the rest of our table may be derived from formula 2. For example,

(2.1)

$$\int_0^\infty e^{-st}\sin at\,dt = \frac{1}{2i}\int_0^\infty e^{-st}[e^{iat} - e^{-iat}]\,dt = \left[\frac{1}{s - ia} - \frac{1}{s + ia}\right]\frac{1}{2i}$$

$$= \frac{a}{s^2 + a^2}.$$

Or one may avoid the use of complex numbers by integrating the integral (2.1) twice by parts:

(2.2)
$$\int_0^\infty e^{-st}\sin at\,dt = \frac{1}{a} - \frac{s}{a}\int_0^\infty e^{-st}\cos at\,dt$$

(2.3)
$$= \frac{1}{a} - \frac{s^2}{a^2}\int_0^\infty e^{-st}\sin at\,dt.$$

Equation (2.3) can be solved to yield formula 3. Note that equation (2.2) then yields also formula 4. The transforms of the hyperbolic functions are obtained similarly. Formula 7 may be obtained from formula 3 by differentiation with respect to s. This operation is justified by the obvious inequality

(2.4)
$$\int_0^\infty e^{-st}t\sin at\,dt \ll \int_0^\infty e^{-\delta t}t\,dt, \qquad 0 < \delta \leqq s < \infty,$$

which shows that the integral (2.4) converges uniformly on (δ, ∞). Similarly, formula 8 follows from formula 3 after differentation with respect to a. [The symbol $\int_a^\infty f(t)\,dt \ll \int_a^\infty g(t)\,dt$ means that $|f(t)| \leq g(t)$ on $a \leq t < \infty$].

3. Solution of Differential Equations

The transform method of solving differential systems will become clear by the perusal of a few special examples. The integration by parts referred to in §1 is displayed in the following formulas:

$$(3.1) \quad \int_0^\infty e^{-st}y'(t)\,dt = -y(0) + s \int_0^\infty e^{-st}y(t)\,dt;$$

$$(3.2) \quad \int_0^\infty e^{-st}y''(t)\,dt = -y'(0) - sy(0) + s^2 \int_0^\infty e^{-st}y(t)\,dt;$$

$$(3.3) \quad \int_0^\infty e^{-st}y'''(t)\,dt = -y''(0) - sy'(0) - s^2 y(0) + s^3 \int_0^\infty e^{-st}y(t)\,dt.$$

The list could be continued in an obvious way, and the conditions on $y(t)$ for the validity of any particular formula are more or less evident. For example, in (3.3) we would assume that $y(t) \in C^3$ on $0 \leq t < \infty$ (is continuous with its first three derivatives) and that $y(t)$, $y'(t)$, and $y''(t)$ are all $O(e^{ct})$, $f \to \infty$, for some c. [$f(x) = O(g(x))$, $x \to \infty$, means that $|f(x)|/g(x)$ is bounded for large x.] Then if either integral (3.3) converges for some s, formula (3.3) will hold, at least for large s.

Example A. Solve

$$y' - y = -2, \qquad y(0) = 1.$$

Set

$$Y(s) = \int_0^\infty e^{-st}y(t)\,dt,$$

and take the Laplace transform of each term of the given differential equation, using (3.1) and formula 1 of the transform table,

$$-1 + sY(s) - Y(s) = -2/s,$$

$$(3.4) \qquad Y(s) = \frac{s-2}{s(s-1)} = \frac{2}{s} - \frac{1}{s-1}.$$

By formulas 1 and 2 of the table we see that the function $2 - e^t$ has the same Laplace transform as $Y(s)$, given by (3.4). Appealing to Theorem 1 for uniqueness we see that

$$(3.5) \qquad\qquad y(t) = 2 - e^t.$$

However, it must not be supposed that we have *proved* that the function (3.5) is the solution of the given system. It really is, and this can now be verified by substitution. What we have proved is that if the system has a solution with the properties which make formula (3.1) valid, then the function (3.5) must be that solution. In actual practice, one does not bother to check such conditions but rather verifies the final solution by substitution. As a matter of fact we shall show in §4 that the method always does give the correct solution, at least for equations of order one.

Example B.

$$y'' + y = \sin t, \qquad y(0) = 0, \quad y'(0) = 0.$$

Now using (3.2) we have

$$s^2 Y + Y = \frac{1}{s^2 + 1},$$

$$Y = \frac{1}{(s^2 + 1)^2}$$

By formula 8 of the transform table and Theorem 1 we obtain as a likely solution

$$(3.6) \qquad\qquad y(t) = \frac{1}{2} \left(\sin t - t \cos t \right).$$

That this is indeed a solution may be verified directly. That it is the only solution satisfying the given boundary conditions follows from the general theory of differential equations.

Example C.

$$y''' + y' = -2 \sin t + 2 \cos t,$$

$$y(0) = 1, \ y'(0) = 0, \ y''(0) = 2.$$

Using (3.1) and (3.3) we proceed as in the preceding example to obtain

$$-2 - s^2 + s^3 Y - 1 + sY = \frac{-2}{s^2 + 1} + \frac{2s}{s^2 + 1},$$

$$Y = \frac{s^4 + 4s^2 + 2s + 1}{s(s^2 + 1)^2} = \frac{2s}{(s^2 + 1)^2} + \frac{2}{(s^2 + 1)^2} + \frac{1}{s}.$$

Again using the table (formulas 4 and 8), we see that Y is the transform of

$$y = t \sin t + \sin t - t \cos t + 1,$$

and this is indeed the desired solution, as is readily checked.

In all these examples we have used the transform to convert the differential system into an algebraic equation whose solution was a rational function. We have then used the method of partial fractions to replace this function by the sum of other rational functions each of which appears as a generating function in our table. Note that we have chosen for entries in this table precisely those generating functions which always appear in the end results after the method of partial fractions is used. Of course, higher powers of $s^2 + a^2$ may also appear, but these could be handled by a more extensive table. (See Exercise 6 at the end of this chapter.) It should thus be clear that the method illustrated by the above example will always work, no matter how high the order of the linear differential equation, provided only that the coefficients of the homogeneous equation are constant and that the boundary conditions bear on a single point (the origin in the above examples). Even if more points are involved the method, slightly modified, may still be used, as we now illustrate.

Example D.

$$y'' + y' = \cos t - \sin t, \quad y(0) = y(\pi) = 0.$$

We assume that the unknown value of $y'(0)$ is the constant A and proceed as in the previous examples.

$$-A + s^2 Y + s Y = \frac{s - 1}{s^2 + 1},$$

$$Y = \frac{1}{s^2 + 1} + \frac{A - 1}{s} + \frac{1 - A}{s + 1},$$

$$y = \sin t + A - 1 + (1 - A)e^{-t}.$$

We have a solution of the differential equation which vanishes at $t = 0$ no matter what the value of A may be. We may now determine A to satisfy the condition $y(\pi) = 0$ and find that $y = \sin t$. Ones sees by inspection that it satisfies all of the required conditions.

4. The Product Theorem

A very useful result, and one which we shall need immediately, is that the product of two or more generating functions is generally a a generating function. In Chapter 5 we shall prove the fact in a more general setting, but let us immediately prove as much as is needed here.

Theorem 4.

$$1. \quad f(s) = \int_0^\infty e^{-st}\varphi(t)\, dt,$$

absolutely convergent at $s = s_0$.

$$2. \quad g(s) = \int_0^\infty e^{-st}\psi(t)\, dt,$$

absolutely convergent at $s = s_0$.

$$(4.1) \quad \Rightarrow \qquad f(s)g(s) = \int_0^\infty e^{-st}\omega(t)\, dt$$

absolutely convergent for $s \geq s_0$,

where

$$\omega(x) = \int_0^x \varphi(t)\psi(x - t)\, dt = \int_0^x \varphi(x - t)\psi(t)\, dt = \varphi * \psi.$$

Recall our agreement in §1 that all determining functions are assumed continuous in the present chapter. Hence the function $\omega(x)$ is well defined for $0 \leq x < \infty$. It is called the *convolution* of φ and ψ. Reverse the order of integration in the iterated integral

$$\int_0^\infty \exp\left(-s_0 x\right) dx \int_0^x \varphi(t)\psi(x - t)\, dt$$

$$= \int_0^\infty \varphi(t)\, dt \int_t^\infty \exp(-s_0 x)\psi(x - t)\, dx$$

$$= \int_0^\infty \varphi(t)\, dt \int_0^\infty \exp[-s_0(t + y)]\psi(y)\, dy = f(s_0)g(s_0).$$

The reversal of order is justified, by Fubini's theorem, if

$$\int_0^\infty |\varphi(t)|\, dt \int_0^\infty \exp[-s_0(t + y)]|\psi(y)|\, dy$$

$$= \int_0^\infty \exp(-s_0 t)|\varphi(t)|\, dt \int_0^\infty \exp(-s_0 y)|\psi(y)|\, dy < \infty.$$

But this inequality holds by the assumed absolute convergence of the two given Laplace integrals. This establishes equation (4.1) at $s = s_0$. But the inequality

$$\int_0^\infty e^{-st}\varphi(t)\, dt \ll \int_0^\infty \exp(-s_0 t)|\varphi(t)|\, dt, \qquad s \geqq s_0$$

shows that absolute convergence of a Laplace integral at one point implies its absolute convergence at all points farther to the right. Thus the proof is complete.

As an example, let us show that $s^{-3}(s^2 + 1)^{-1}$ is a generating function. From our table we have

$$\frac{1}{s^3} = \int_0^\infty e^{-st} \frac{t^2}{2}\, dt,$$

$$\frac{1}{s^2 + 1} = \int_0^\infty e^{-st} \sin t\, dt$$

$$\omega(x) = \frac{1}{2}\int_0^x (x - t)^2 \sin t\, dt = \frac{x^2}{2} + \cos x - 1,$$

$$\frac{1}{s^3(s^2 + 1)} = \int_0^\infty e^{-st}\left(\frac{t^2}{2} - 1 + \cos t\right) dt.$$

This result could be checked by the use of partial fractions:

$$\frac{1}{s^3(s^2+1)} = \frac{1}{s^3} + \frac{s}{s^2+1} - \frac{1}{s}.$$

Now, let us show that the method is always valid, without checks, for the general system

$$(4.2) \qquad\qquad y' + ay = \varphi(t), \qquad y(0) = A$$

provided only that $\varphi(x)$ is the determining function of an absolutely convergent Laplace transform:

$$f(s) = \int_0^\infty e^{-st}\varphi(t)\,dt, \qquad \text{absolutely convergent at } s_0.$$

Proceeding as usual,

$$-A + sY + aY = f(s)$$

$$Y = \frac{A+f(s)}{s+a} = \frac{A}{s+a} + \frac{f(s)}{s+a}.$$

Now $A/(s+a)$ is the transform of Ae^{-at} and $f(s)/(s+a)$, of

$$\omega(t) = \int_0^t e^{-a(t-y)}\varphi(y)\,dy,$$

by Theorem 4. Hence by Theorem 1,

$$(4.3) \qquad\qquad y(t) = Ae^{-at} + \omega(t).$$

But now we can show once and for all that this is the solution of the given system. For,

$$(y(t)e^{at})' = \frac{d}{dt}\int_0^t e^{ay}\varphi(y)\,dy = e^{at}\varphi(t).$$

But this is clearly equivalent to the differential equation (4.2). Since $\omega(0) = 0$, it is also evident that $y(0) = A$.

A similar proof is possible for equations of higher order. See D. V. Widder (1961, p. 482). The purpose of the above presentation is not the solution of a trivial differential system, but is rather to suggest the scope and the limitations of the method. Note that the solution would be impossible by use of transforms as outlined above if $\varphi(t) = \exp t^2$, and yet the function of (4.3) is the solution of (4.2), even in this case. For, the function $\omega = e^{-at} * e^{t^2}$ is well defined and satisfies (4.2).

5. Integral Equations

An integral equation is one in which an unknown function appears under an integral sign. Such an equation can sometimes be reduced, by differentiation, to a differential equation. But in most cases such a reduction is not possible, and independent methods of solution must be devised. If the integral involved is a convolution, as defined in §4, then the solution may often be found conveniently by use of the Laplace transform. Our first illustration will be one in which the equation can be reduced to a differential equation, so that the result can be checked.

Example A. Solve for $y(t)$ the integral equation

$$(5.1) \qquad y(t) = t - \sin t - \int_0^t (t - z)y(z)\, dz.$$

Note that no boundary condition is needed. Set $Y(s)$ equal to the transform of the unknown function $y(t)$, use the table and Theorem 4 to obtain

$$Y = \frac{1}{s^2} - \frac{1}{s^2 + 1} - \frac{Y}{s^2}$$

$$Y = \frac{1}{(s^2 + 1)^2}.$$

Again use the table and Theorem 1 to find that

$$(5.2) \qquad y(t) = \frac{1}{2}(\sin t - t \cos t).$$

As in the case of differential equations, this result should now be checked by direct substitution in (5.1). However, in this case another check is available. Two differentiations reduce (5.1) to a differential equation:

$$(5.3) \qquad y' = 1 - \cos t - \int_0^t y(z)\, dz,$$

$$y'' + y = \sin t.$$

From (5.1) and (5.3) it is clear that $y(0) = y'(0) = 0$. But this system was solved as Example B in §3, and (5.2) was the solution.

Example B.

$$y'(t) = \cos t + \int_0^t y(t - z) \cos z\, dz, \qquad y(0) = 1.$$

This is called an integrodifferential equation, since the unknown function is involved both in an integral and in a derivative. Note the boundary condition, the need for which will become apparent. Using (3.1), the table, and Theorem 4 we have

$$-1 + sY = \frac{s}{s^2 + 1} + \frac{Y}{s^2 + 1}$$

$$Y = \frac{1}{s} + \frac{1}{s^2} + \frac{1}{s^3}$$

$$y = 1 + t + \frac{t^2}{2}.$$

The solution should be checked by substitution. But see Exercise 8 of this chapter for an alternative procedure.

Example C.

$$\varphi(t) = \int_0^t \frac{y(z)\, dz}{(t - z)^\alpha}, \qquad 0 < \alpha < 1.$$

This is known as Abel's integral equation. See N. H. Abel [1826]. Here $y(t)$ is to be found and $\varphi(t)$ is a given function which we assume to have a derivative which is continuous on $0 \le t < \infty$. Clearly $\varphi(0)$ must be zero from the original equation. Let us assume further that

$$(5.4) \qquad \varphi(t) = O(e^{at}), \qquad \text{some } a, \quad t \to +\infty,$$

so that the following integral converges absolutely for $s > a$ and defines a function $f(s)$,

$$(s) = \int_0^\infty e^{-st} \varphi(t)\, dt,$$

$$sf(s) = \int_0^\infty e^{-st} \varphi'(t)\, dt.$$

In the integration by parts used here, the integrated part vanishes by virtue of (5.4). Defining $Y(s)$ as the transform of $y(t)$, Theorem 4 yields

$$f(s) = \Gamma(1 - \alpha)s^{\alpha - 1} Y(s)$$

$$(5.5) \qquad Y(s) = \frac{s f(s)}{\Gamma(1 - \alpha)s^\alpha}$$

Another application of Theorem 4 to the product (5.5) shows that $Y(s)$ is the transform of

$$\varphi'(t) * t^{\alpha - 1}/(\Gamma(1 - \alpha)\Gamma(\alpha)).$$

By Theorem 1

$$(5.6) \qquad y(t) = \frac{\sin \pi\alpha}{\pi} \int_0^t \frac{\varphi'(z)}{(t - z)^{1 - \alpha}}\, dz.$$

We can now show by direct substitution that this function does indeed satisfy Abel's integral equation.

$$\int_0^t \frac{y(z)}{(t - z)^\alpha}\, dz = \frac{\sin \pi\alpha}{\pi} \int_0^t \frac{dz}{(t - z)^\alpha} \int_0^z \frac{\varphi'(r)}{(z - r)^{1 - \alpha}}\, dr$$

$$= \frac{\sin \pi\alpha}{\pi} \int_0^t \varphi'(r)\, dr \int_r^t \frac{dz}{(t - z)^\alpha (z - r)^{1 - \alpha}}$$

$$= \frac{\sin \pi\alpha}{\pi} \int_0^t B(1 - \alpha, \alpha)\varphi'(r)\, dr = \varphi(t).$$

Here we have used the familiar fact that

$$\int_r^t \frac{dz}{(t - z)^\alpha (z - r)^{1 - \alpha}} = \int_0^1 r^{-\alpha}(1 - r)^{\alpha - 1}\, dr = B(1 - \alpha, \alpha) = \frac{\pi}{\sin \pi\alpha}.$$

Example D. (Abel's mechanical problem). The above integral equation arises naturally in the following problem. A wire in the form of a plane curve, $y = f(x)$, through the origin of an xy-plane is fixed so that the positive y-axis extends vertically upwards in the earth's gravitational field. A frictionless bead slides down the wire starting from rest at an arbitrary point (a, b). Determine the shape of the curve so that the time for the bead to reach the origin shall be a prescribed function $\varphi(b)$ of the starting height. It is known, and easily proved, that the bead's

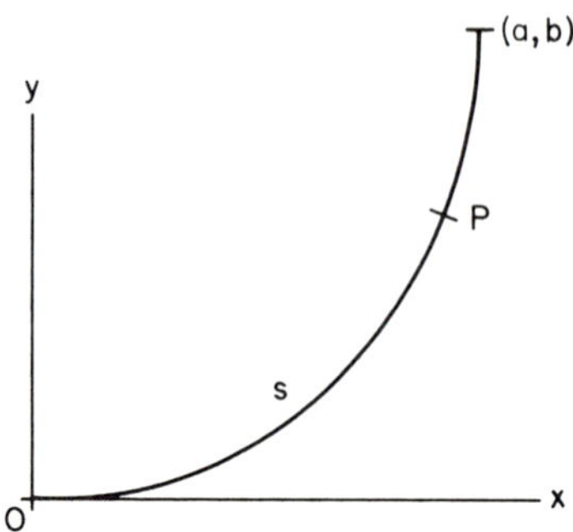

velocity at a point P with coordinates (x, y) is the same as if it had fallen vertically through the distance $b - y$. This velocity v is $[2g(b - y)]^{1/2}$, where g is the acceleration of gravity. The total time of descent is obtained from this by integration:

$$\varphi(b) = \int \frac{ds}{v} = \frac{1}{(2g)^{1/2}} \int_0^b \frac{\dfrac{ds}{dy}}{(b - y)^{1/2}}\, dy.$$

But this is Abel's equation with $\alpha = 1/2$. Its solution (5.6) is

$$s'(y) = \frac{(2g)^{1/2}}{\pi} \int_0^y \frac{\varphi'(z)}{(y - z)^{1/2}}\, dz = \frac{(2g)^{1/2}}{\pi} \frac{d}{dy} \int_0^y \frac{\varphi(y - z)}{\sqrt{z}}\, dz$$

$$(5.7) \quad s(y) = \frac{(2g)^{1/2}}{\pi} \int_0^y \frac{\varphi(y - z)}{\sqrt{z}}\, dz.$$

Then from the relation $ds^2 = dx^2 + dy^2$ one can determine the equation of the desired curve by integration:

$$x = \int_0^y ([s'(y)]^2 - 1)^{1/2}\, dy.$$

For example, if $\varphi(b) = \pi[b/(2g)]^{1/2}$ then $s(y) = \pi y/2$ and the desired curve is the straight line

$$x = y\left[\frac{\pi^2}{4} - 1\right]^{1/2}.$$

Example E. Determine the curve of the previous example in such a way that the time of descent of the bead shall always be the same, regardless of the starting point of the wire. That is, $\varphi(b)$ must be independent of b. Integrate (5.7) to obtain

$$(5.8) \qquad s(y) = \frac{\varphi}{\pi}(8gy)^{1/2}, \qquad \varphi = \varphi(b).$$

From the constancy of the time of descent the curve defined by (5.8) is called the *tautochrone* (same time). Let us show that it is a cycloid. Since the constant φ is unimportant for the shape of the curve let us take it equal to π. Then

$$\frac{ds}{dy} = \frac{(2g)^{1/2}}{\sqrt{y}}, \qquad \frac{dx}{ds} = \left(1 - \frac{y}{2g}\right)^{1/2} = \left(1 - \frac{s^2}{16g^2}\right)^{1/2},$$

$$x = \int\left(1 - \frac{s^2}{16g^2}\right)^{1/2} ds = 2g\int \cos^2\frac{\theta}{2}\, d\theta, \qquad s = 4g\sin\frac{\theta}{2}.$$

Determine the constant of integration so that $\theta = s = x = 0$. Then

$$(5.9) \qquad \begin{aligned} x &= g(\theta + \sin\theta), \\ y &= g(1 - \cos\theta). \end{aligned}$$

This pair of parametric equations defines a cycloid. It is traced by that point of a circle of radius g which was originally at the origin, as the circle rolls on the line $y = 2g$ on the under side of it.

EXERCISES

1. Solve $y' - y = \cos t$, $\qquad y(0) = -1$.

2. Solve $y' - 2y = 2t^2 - 2t$ $\qquad y(0) = 1$.

3. Solve and check

$$y'' - y = t, \qquad y(0) = 0, \quad y'(0) = 1.$$

4. Solve and check

$$y'' + 2y' + y = 1, \qquad y(0) = y'(0) = 0.$$

5. Solve and check

$$y''' - 4y' = \cosh 2t - \sinh 2t,$$

$$y(0) = 1, \quad y'(0) = -6, \quad y''(0) = 4.$$

$$\text{Ans.} \quad \cosh 2t - 3 \sinh 2t.$$

6. Show that the Laplace transform of $t \sin at - at^2 \cos at$ is $8a^3 s(s^2 + a^2)^{-3}$, thus enlarging the table of §2.

7. Solve and check

$$y^{(4)} + 2y'' + y = 8 \cos t, \qquad y(0) = y'(0) = y''(0) = y'''(0) = 0.$$

8. Reduce the integral equation of Example B, §5, to a differential system and solve.

9. Solve by two methods and check

$$y(t) = 2t + \int_0^t e^{t-z} y(z)\, dz.$$

10. Show that under the assumptions made about $\varphi(t)$ in Example C of §5 the solution (5.6) may take the form

$$y(t) = \frac{\sin \pi \alpha}{\pi} \frac{d}{dt} \int_0^t \frac{\varphi(z)}{(t - z)^{1-\alpha}}\, dz.$$

11. Solve Abel's equation when the given function is t.

12. If $\alpha > 0$ define the fractional integral $D^{-\alpha}$ of order α of a function $f(x)$ as

$$D^{-\alpha} f(x) = \frac{1}{\Gamma(\alpha)} \int_0^x (x - t)^{\alpha-1} f(t)\, dt.$$

For $0 < \alpha < 1$ define the fractional derivative D^α of order α of a function $f(x)$ as

$$D^\alpha f(x) = \frac{d}{dx} D^{\alpha-1} f(x).$$

Prove

$$D^{-\alpha}x^p = \frac{\Gamma(p+1)}{\Gamma(p+\alpha+1)}\, x^{p+\alpha} \qquad \alpha > 0, \quad p \geqq 0$$

$$D^{\alpha}x^p = \frac{\Gamma(p+1)}{\Gamma(p+1-\alpha)}\, x^{p-\alpha}, \qquad 0 < \alpha < 1, \quad p \geqq 0.$$

13. Solve the fractional integral equation

$$D^{-\alpha}f(x) = \varphi(x), \qquad 0 < \alpha < 1.$$

What are you assuming about $\varphi(x)$?

14. Solve the fractional differential equation

$$D^{\alpha}f(x) = \varphi(x), \qquad 0 < \alpha < 1.$$

Is the solution unique?

15. Prove in Example D of §5 that the velocity of the bead at (x, y) is the same as if it had fallen vertically to that level from (a, b). Resolve the force of gravity in the direction of the tangent to the curve and use Newton's second law of motion.

16. Plot the curve (5.9). Compute the arc length from $(0,0)$ to (a, b). Describe the geometric meaning of the parameter θ.

2 *Dirichlet Series*

1. Introduction

Three cases of *transforms* of frequent occurrence in analysis are:

A.
$$f(s) = \int_0^\infty G(s, t)\varphi(t)\, dt,$$

B.
$$f_n = \sum_{k=0}^\infty G(n, k)\varphi_k,$$

C.
$$f(s) = \sum_{k=0}^\infty G(s, k)\varphi_k.$$

Here, and later throughout this book, s is to be a complex variable, $s = \sigma + i\tau$. The above operations convert one function or sequence into another. Familiar examples occur, for instance, in the following theories.

    I. The Laplace transform: $G(s, t) = e^{-st}$;

19

II.   Cesàro summability: $G(n, k) = \dfrac{1}{n}, \qquad k \leqq n$

$$= 0, \qquad k > n;$$

III.   Power series: $G(s, k) = s^k.$

We shall begin our studies in the present chapter with the example of Dirichlet series from case C. The theory involved will be simple in the sense that much of it could be conjectured from the theory of power series. And yet this "discrete" transform will provide a sort of model for the more complicated integral transforms to follow.

We define $G(s, k)$ as $\exp(-\lambda_k s)$ where $0 \leqq \lambda_1 < \lambda_2 < \ldots$ and $\lambda_k \to \infty$ as $k \to \infty$. At least when $\lambda_k = k$ the region of convergence must be a right half-plane,

$$|e^{-s}| < \rho \quad \text{or} \quad \sigma > \log \frac{1}{\rho},$$

since it then becomes a power series in e^{-s}. One might thus conjecture that this property holds in general. Similarly, we might predict that the sum of a Dirichlet series is analytic, that its coefficients are uniquely determined by the sum and that they can be found, as in the case of power series, by a contour integration. All these conjectures will be established. By contrast we shall find that the sum of a general Dirichlet series need have no singularity on the axis of convergence.

2. Convergence Tests

Two tests that are of frequent use for the convergence of a Dirichlet series are due to Cauchy. The first is the familiar *integral test* which we state without proof.

Theorem 2.1.   1. $f(x) \in C, \downarrow, \geqq 0, \qquad a \leqq x < \infty.$

$\Rightarrow$ The series and integral

$$\sum_{k > a}^{\infty} f(k), \qquad \int_{a}^{\infty} f(x)\, dx$$

converge or diverge together.

The second is known as the *condensation test.*

Theorem 2.2.

1. $$f(x) \in C, \downarrow, \geqq 0, \qquad a < x < \infty,$$

2. $$b > 1.$$

$\Rightarrow$ The two series

$$\sum_{k>a}^{\infty} f(k), \qquad \sum_{k>a}^{\infty} b^k f(b^k)$$

converge or diverge together.

This test is an extension of the familiar trick of grouping terms in the harmonic series

$$1 + \left(\frac{1}{2} + \frac{1}{3}\right) + \left(\frac{1}{4} + \frac{1}{5} + \frac{1}{6} + \frac{1}{7}\right) + \cdots \gg 1 + \frac{2}{4} + \frac{4}{8} + \cdots$$

to show its divergence. The word " condensation " is used since one need only consider the behavior of a condensed portion of the series, in this example the 2^n terms in the nth parenthesis, in order to prove its divergence.

Since $b^x \in \uparrow, f \geqq 0$, and $f(b^x) \in \downarrow$, we have

$$b^k f(b^{k+1}) \leqq b^x f(b^x) \leqq b^{k+1} f(b^k), \qquad k \leqq x \leqq k + 1$$

$$\frac{1}{b} \sum_{k=m+1}^{n+1} b^k f(b^k) \leqq \int_m^{n+1} b^x f(b^x)\, dx \leqq b \sum_{k=m}^{n} b^k f(b^k) \qquad n > m \geqq a.$$

From these inequalities it is clear that the series and integral

$$\sum_m^{\infty} b^k f(b^k), \qquad \int_m^{\infty} b^x f(b^x)\, dx = \frac{1}{\log b} \int_{b^m}^{\infty} f(x)\, dx$$

converge and diverge together. But by Theorem 2.1 the latter integral will converge if and only if the series $\sum^{\infty} f(k)$ does. This concludes the proof.

Example A. The series

$$\sum_{k=2}^{\infty} \frac{1}{k(\log k)^p}$$

converges for $p > 1$, diverges for $p = 1$. For, by Theorem 2.2, it has the same convergence properties as the more familiar series

$$\frac{1}{(\log 2)^p} \sum_{k=1}^{\infty} \frac{1}{k^p}.$$

3. Convergence of Dirichlet Series

Let us begin with a formal definition.

Definition 3. If $0 \leqq \lambda_1 < \lambda_2 < \lambda_3 < \ldots$, and if $\lim \lambda_u = \infty$, the series

$$(3.1) \qquad \sum_{k=1}^{\infty} a_k \exp(-\lambda_k s)$$

is called a Dirichlet series of type λ_k.

If the type is $\lambda_k = \log k$, the series is also called *ordinary*:

$$(3.2) \qquad \sum_{k=1}^{\infty} \frac{a_k}{k^s}.$$

This is the type originally used by Dirichlet for his studies in number theory and includes the familiar series for the Riemann zeta-function,

$$\zeta(s) = \sum_{k=1}^{\infty} \frac{1}{k^s}.$$

If the type is $\lambda_k = k$, the series (3.1) becomes a power series. It is sometimes useful to note that series (3.2) also may reduce to a power series— for example, if $\lambda_k = k \log 2$ in (3.1).

We shall have frequent use for the following lemma.

Lemma 3.

$$1. \quad \left| \sum_{k=1}^{n} a_k \exp(-\lambda_k s_0) \right| \leqq M, \qquad n = 1, 2, 3, \cdots;$$

$$2. \quad \sigma > \sigma_0$$

$$\Rightarrow \quad \left| \sum_{k=1}^{\infty} a_k \exp(-\lambda_k s) \right| \leqq M \, \frac{|s - s_0|}{\sigma - \sigma_0} \exp[-\lambda_1(\sigma - \sigma_0)].$$

It is of course assumed here that the sequence $\{\lambda_k\}^\infty$ satisfies the conditions of Definition 3. Set

$$U_n = \sum_{k=1}^{n} a_k \exp(-\lambda_k s_0).$$

Then

$$(3.3) \quad \sum_{k=1}^{n} a_k \exp(-\lambda_k s) = U_1 \exp[-\lambda_1(s - s_0)]$$

$$+ \sum_{k=2}^{n} (U_k - U_{k-1}) \exp[-\lambda_k(s - s_0)]$$

$$(3.4) \qquad = \sum_{k=1}^{n-1} U_k[\exp(-\lambda_k(s-s_0)) - \exp(-\lambda_{k+1}(s-s_0))]$$

$$+ U_n \exp[-\lambda_n(s - s_0)].$$

But the last term tends to zero with $1/n$ and

$$\exp[-\lambda_k(s - s_0)] - \exp[-\lambda_{k+1}(s - s_0)]$$

$$= (s - s_0) \int_{\lambda_k}^{\lambda_{k+1}} \exp[-t(s - s_0)]\, dt,$$

so that

$$(3.5) \quad \sum_{k=1}^{\infty} a_k \exp(-\lambda_k s) = (s - s_0) \sum_{k=1}^{\infty} U_k \int_{\lambda_k}^{\lambda_{k+1}} \exp[-t(s - s_0)]\, dt$$

provided either series converges. But the latter is absolutely convergent since

$$\sum_{k=1}^{\infty} U_k \int_{\lambda_k}^{\lambda_{k+1}} \exp[-t(s - s_0)]\, dt \ll M \sum_{k=1}^{\infty} \int_{\lambda_k}^{\lambda_{k+1}} \exp[-t(\sigma - \sigma_0)]\, dt$$

$$(3.6) \qquad = M \int_{\lambda_1}^{\infty} \exp[-t(\sigma - \sigma_0)]\, dt = \frac{M}{\sigma - \sigma_0} \exp[-\lambda_1(\sigma - \sigma_0)].$$

Now the conclusion of the lemma follows from (3.5) and (3.6).

The algebra which converts equation (3.3) into equation (3.4) is known as *partial summation*. It here has the effect of equating a series which may be conditionally convergent to an absolutely convergent one [Equation (3.5)].

One trivial consequence of the lemma is that the sum of a Dirichlet series is uniformly bounded in any horizontal strip of finite width *inside* the region of convergence;

$$\sigma \geqq \sigma_c + \delta, \ |\tau| \leqq R \qquad \text{some } \delta > 0, \quad R > 0.$$

We can now prove that the region of convergence of a Dirichlet series is in general a half-plane.

Theorem 3.1.

$$(3.7) \qquad 1. \quad \sum_{k=1}^{\infty} a_k \exp(-\lambda_k s) \qquad \text{converges at } s_0$$

$$\Rightarrow \qquad \text{it converges in the half-plane } \sigma > \sigma_0.$$

Since the partial sums of a convergent series are bounded, hypothesis 1 of Lemma 3 holds, and the conclusion of Theorem 3.1 follows.

As a corollary it is evident that if (3.7) diverges at s_0 it does so for $\sigma < \sigma_0$. By use of the Dedekind cut it follows that (3.7) must converge for all s, for no s, or there must exist a vertical line $\sigma = \sigma_c$ which divides the right half-plane of convergence from the left half-plane of divergence. The number σ_c is called the *abscissa of convergence*. The notations $\sigma_c = +\infty$, $\sigma_c = -\infty$ are admitted for series which converge nowhere and everywhere, respectively. That all three cases may occur may be seen from the known facts about power series ($\lambda_k = k$) or from the examples

$$\sum_{k=1}^{\infty} \frac{k!}{k^s}, \qquad \sum_{k=1}^{\infty} \frac{1}{k^s}, \qquad \sum_{k=1}^{\infty} \frac{1}{k!\,k^s},$$

for which $\sigma_c = +\infty$, $\sigma_c = 1$, and $\sigma_c = -\infty$, respectively.

The absolute convergence of a Dirichlet series is treated more simply.

Theorem 3.2.

$$(3.8) \qquad 1. \quad \sum_{k=1}^{\infty} a_k \exp(-\lambda_k s) \qquad \text{converges absolutely at } s_0$$

$$\Rightarrow \qquad \text{it converges absolutely for } \sigma > \sigma_0.$$

This follows immediately from the relation

$$\sum_{k=1}^{\infty} a_k \exp(-\lambda_k s) \ll \sum_{k=1}^{\infty} |a_k| \exp(-\lambda_k \sigma_0), \qquad \sigma > \sigma_0.$$

As before we can now show the existence of an *abscissa of absolute convergence* σ_a, which may be $\pm\infty$, as a result of Theorem 3.2. That σ_a need not equal σ_c may be seen from the example

$$(3.9) \qquad \eta(s) = \sum_{k=1}^{\infty} \frac{(-1)^{k+1}}{k^s}, \qquad \sigma_c = 0, \quad \sigma_a = 1.$$

One consequence of Theorems 3.1 and 3.2 is that we may restrict attention to real s when determining σ_a and σ_c. Dirichlet's test for alternating series shows that (3.9) converges for real positive s. Since (3.9) diverges for $s = 0$, $\sigma_c = 0$. On the other hand $\sigma_a = 1$ since

$$\sum_{k=1}^{\infty} \left| \frac{(-1)^{k+1}}{k^s} \right| = \sum_{k=1}^{\infty} \frac{1}{k^\sigma} = \zeta(\sigma).$$

From the classical theory of power series we know that $\sigma_c = \sigma_a$ when $\lambda_k = k$. But if no restriction is placed on the λ_k there is no upper bound for the difference $\sigma_a - \sigma_c$. (See Exercise 3 at the end of this chapter.) Compare also Theorem 6.2, below.

4. Analyticity

We can now show easily that a Dirichlet series represents an analytic function in its region of convergence.

Theorem 4.

$$1. \quad f(s) = \sum_{k=1}^{\infty} a_k \exp(-\lambda_k s);$$

$$2. \quad \sigma_c < +\infty$$

$\Rightarrow$ A. $f(s) \in A$ (is analytic), $\qquad \sigma > \sigma_c$.

$$\text{B.} \quad f^{(p)}(s) = \sum_{k=1}^{\infty} (-\lambda_k)^p a_k \exp(-\lambda_k s),$$

$$p = 1,2,3,\ldots, \sigma > \sigma_c.$$

To prove this we appeal to the classical theorem of Weierstrass concerning series of analytic function. Let s_1 be an arbitrary point in the half-plane $\sigma > \sigma_c$, and let D be a disc of radius ρ with s_1 as center and also entirely in that half-plane. We show that the series

$$(4.1) \qquad \sum_{k=1}^{\infty} a_k \exp(-\lambda_k s)$$

converges uniformly in D. That is, for each $\varepsilon > 0$ there is an N, independent of s in D, such that for $n > N$

$$(4.2) \qquad \left| \sum_{k=n}^{\infty} a_k \exp(-\lambda_k s) \right| < \varepsilon, \qquad s \in D.$$

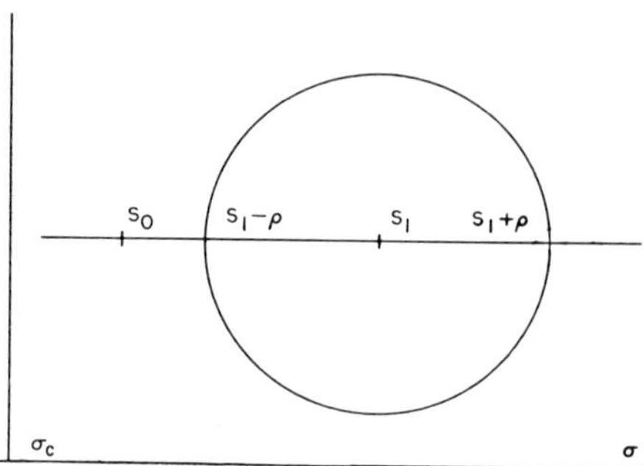

Figure 1

We apply Lemma 3 to this truncated series choosing s_0 with the same ordinate as s_1 and such that $\sigma_c < \sigma_0 < \sigma_0 + \delta = \sigma_1 - \rho$. Since (4.1) converges at s_0 the partial sums of (4.1), at s_0, are bounded in absolute value by a number M. The corresponding partial sums for (4.2) are then bounded by $2M$,

$$\left| \sum_{n}^{n+k} a_k \exp(-\lambda_k s_0) \right| = \left| \sum_{1}^{n+k} a_k \exp(-\lambda_k s_0) - \sum_{1}^{n-1} a_k \exp(-\lambda_k s_0) \right| \leq 2M,$$

and the lemma gives

$$\left| \sum_{n}^{\infty} a_k \exp(-\lambda_k s) \right| \leq 2M \, \frac{s - s_0}{\sigma - \sigma_0} \exp[-\lambda_n(\sigma - \sigma_0)], \qquad \sigma > \sigma_0,$$

$$\leq 2M \, \frac{\sigma_1 + \rho - \sigma_0}{\delta} \exp(-\lambda_n \delta), \qquad s \in D.$$

The right-hand side is independent of s and tends to zero as $n \to \infty$, so that the existence of N for (4.2) is assured. By the Weierstrass theorem $f(s)$ is analytic in D, and term-by-term differentiation is permissibly there. The conclusions of the theorem are thus established for the *arbitrary* point s_1.

5. Uniform Convergence

It is clear from the previous section that a Dirichlet series converges uniformly in any compact region of its half-plane of convergence. It also converges uniformly in certain regions which extend to infinity. Let us introduce a definition and a notation.

Definition 5.1. A Stolz region for the point s_0, denoted by $\mathrm{St}(s_0)$, is the set

$$\{s\,|\,|\arg\,(s - s_0)| \leqq a,\ \sigma \geqq \sigma_0\}$$

for some $a < \pi/2$.

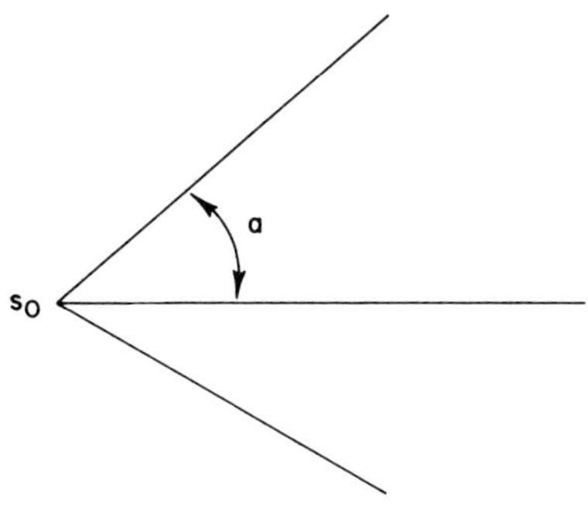

Figure 2

This is an angular region lying between two intersecting lines, as indicated in Fig. 2. The angle a is not indicated in the notation since it may usually be arbitrary, less than $\pi/2$.

Theorem 5.

$$(5.1) \qquad 1. \quad \sum_{k=1}^{\infty} a_k \exp(-\lambda_k s) \qquad \text{converges at } s_0$$

$$\Rightarrow \qquad \text{it converges uniformly in } \mathrm{St}(s_0).$$

We must show that for each $\varepsilon > 0$ there is an N, independent of s in $St(s_0)$ such that when $n > N$

$$(5.2) \qquad \left| \sum_{n}^{\infty} a_k \exp(-\lambda_k s) \right| \le \varepsilon, \qquad s \in St(s_0).$$

By the Cauchy criterion for the existence of a limit we may choose N so that

$$(5.3) \qquad \left| \sum_{n}^{n+p} a_k \exp(-\lambda_k s_0) \right| \le \varepsilon \cos a, \qquad n > N, \quad p = 1, 2, \ldots,$$

since (5.1) converges at s_0. But these are the partial sums of a truncated Dirichlet series with first term $a_n \exp(-\lambda_n s_0)$ to which we may now apply Lemma 3 to obtain

$$\left| \sum_{n}^{\infty} a_k \exp(-\lambda_k s) \right| \le \varepsilon \cos a \, \frac{|s - s_0|}{\sigma - \sigma_0} \exp[-\lambda_n(\sigma - \sigma_0)], \qquad \sigma > \sigma_0.$$

But in $St(s_0)$, when $s \ne s_0$,

$$\frac{|s - s_0|}{\sigma - \sigma_0} \le \sec a, \qquad \exp[-\lambda_n(\sigma - \sigma_0)] < 1.$$

Hence (5.2) is proved when $s \ne s_0$. Allowing p to become infinite in (5.3) we also have the result at s_0.

If we denote the sum of (5.1) by $f(s)$, we have under the hypothesis of Theorem 5 that

$$(5.4) \qquad \lim_{s \to s_0} f(s) = f(s_0),$$

$$(5.5) \qquad \lim_{s \to \infty} f(s) = \begin{cases} 0, & \lambda_1 > 0, \\ a_1, & \lambda_1 = 0, \end{cases}$$

provided that s remains inside $St(s_0)$ as it approaches its limit. In particular

$$(5.6) \qquad \lim_{\sigma \to \infty} \exp(\lambda_1 \sigma) f(\sigma) = a_1,$$

whether λ_1 is zero or not. Of course (5.4) is also a result of Theorem 4 when s_0 is off the axis of convergence. If s_0 is on that axis, (5.4) is the analog of the classic theorem of Abel for power series.

6. Formulas for σ_c and σ_a

The familiar formula for the radius of convergence ρ of a power series with coefficients a_n,

$$\frac{1}{\rho} = \overline{\lim_{n \to \infty}} \, |a_n|^{1/n},$$

does not generalize in the most obvious manner to Dirichlet series:

$$\sigma_c = \log \frac{1}{\rho} = \overline{\lim_{n \to \infty}} \, \frac{\log |a_n|}{\lambda_n}.$$

This clearly fails for the ζ-function, for which it would give 0 rather than the correct value 1. Further reflection shows that it could not be a general result, since it would always make $\sigma_a = \sigma_c$, involving as it does the absolute values of the coefficients. To arrive at a correct conjecture note that the two series

$$\sum_{k=0}^{\infty} a_k z^k, \qquad \sum_{k=0}^{\infty} U_k z^k = \frac{1}{1-z} \sum_{k=0}^{\infty} a_k z^k,$$

where

$$U_n = \sum_{k=0}^{n} a_k,$$

have the same radius of convergence ρ, at least if it is less than 1:

$$\frac{1}{\rho} = \overline{\lim_{n \to \infty}} |U_n|^{1/n}.$$

Taking logarithms of both sides and replacing n by λ_n one arrives at the correct formula, due to E. Cahen [1894]:

$$\sigma_c = \overline{\lim_{n \to \infty}} \, \frac{\log |U_n|}{\lambda_n}, \qquad \sigma_c > 0.$$

To prove Cahen's formula we use two lemmas.

Lemma 6.1.

$$1. \quad U_n = \sum_{k=1}^{n} a_k = O\,(\exp \alpha\lambda_n), \qquad n \to \infty\,;$$

$$2. \quad \alpha > 0$$

$$\Rightarrow \qquad \sum_{k=1}^{\infty} a_k \exp(-\lambda_k s) \quad \text{converges for } \sigma > \alpha.$$

By (3.4) with $s_0 = 0$

$$\sum_{k=1}^{n} a_k \exp(-\lambda_k s)$$

$$= \sum_{k=1}^{n-1} U_k [\exp(-\lambda_k s) - \exp(-\lambda_{k+1}s)] + U_n \exp(-\lambda_n s).$$

By hypothesis 1,

$$U_n \exp(-\lambda_n s) = o(1), \qquad n \to \infty, \quad \sigma > \alpha.$$

Hence we need only show that

$$\sum_{k=1}^{\infty} U_k \int_{\lambda_k}^{\lambda_{k+1}} e^{-st}\, dt$$

converges for $\sigma > \alpha$. But it is dominated by the series

$$(6.1) \quad M \sum_{k=1}^{\infty} \int_{\lambda_k}^{\lambda_{k+1}} \exp[-t(\sigma - \alpha)]\, dt = M \int_{\lambda_1}^{\infty} \exp[-t(\sigma - \alpha)]\, dt,$$

$$\text{some } M.$$

Here we have used the assumed rate of growth of U_n and the fact that for $\alpha > 0$

$$\exp \lambda_k \alpha < \exp t\alpha, \qquad \lambda_k < t.$$

Since the integral (6.1) converges for $\sigma > \alpha$, the proof is complete.
The following result is a partial converse to the above.

Lemma 6.2.

$$1. \quad \sum_{k=1}^{\infty} a_k \exp(-\lambda_k \alpha) \quad \text{converges}, \qquad \text{some } \alpha > 0$$

$$\Rightarrow \qquad U_n = \sum_{k=1}^{n} a_k = O(\exp \alpha\lambda_n), \qquad n \to \infty.$$

The partial sums of the series in hypothesis 1 are bounded:

$$|V_n| = \left| \sum_{k=1}^{n} a_k \exp(-\lambda_k \alpha) \right| \leqq M, \qquad n = 1, 2, \ldots .$$

But

$$U_n = \sum_{k=1}^{n} a_k \exp(-\lambda_k \alpha) \exp \lambda_k \alpha$$

$$= V_1 \exp \lambda_1 \alpha + \sum_{k=2}^{n} (V_k - V_{k-1}) \exp \lambda_k \alpha$$

$$= \sum_{k=1}^{n-1} V_k[\exp \lambda_k \alpha - \exp \lambda_{k+1}\alpha] + V_n \exp \lambda_n \alpha$$

$$|U_n| \leqq M \sum_{k=1}^{n-1} [\exp \lambda_{k+1}\alpha - \exp \lambda_k \alpha] + M \exp \lambda_n \alpha$$

$$= M[-\exp \lambda_1\alpha + 2 \exp \lambda_n \alpha] = O(\exp \lambda_n \alpha), \qquad n \to \infty.$$

This completes the proof. Notice where it would fail if $\alpha < 0$.
We now prove Cahen's formula.

Theorem 6.1.

$$1. \quad \overline{\lim_{n \to \infty}} \ \frac{\log |\sum_{k=1}^{n} a_k|}{\lambda_n} = \alpha \quad (\text{or } +\infty);$$

$$2. \quad \alpha > 0$$

$$(6.2) \quad \Rightarrow \quad \sigma_c \ \text{for} \ \sum_{k=1}^{\infty} a_k \exp(-\lambda_k s) \ \text{is } \alpha \quad (\text{or } +\infty).$$

Let us assume that α is finite and prove that series (6.2) converges for $\sigma > \alpha$. If $\varepsilon > 0$, then hypothesis 1 implies that

$$\sum_{k=1}^{n} a_k = O[\exp(\alpha + \varepsilon)\lambda_n], \qquad n \to \infty.$$

By Lemma 6.1. it follows that (6.2) converges for $\sigma > \alpha + \varepsilon$, every $\varepsilon > 0$, and hence for $\sigma > \alpha$.

We show next that (6.2) diverges for $\sigma < \alpha$. Suppose the contrary. Then there exists a positive number $\beta < \alpha$ such that (6.2) converges for $s = \beta$. By Lemma 6.2

$$\left| \sum_{k=1}^{n} a_k \right| < M \exp \lambda_n \beta, \quad \text{some } M, \quad n = 1, 2, 3, \ldots$$

$$\frac{\log |\sum_{k=1}^{n} a_k|}{\lambda_n} < \frac{\log M}{\lambda_n} + \beta.$$

The limit superior of the left-hand side is α, by hypothesis 1, so that $\alpha \leq \beta$. This contradicts the assumption $\beta < \alpha$. The proof for $\alpha = \infty$ is left to the reader.

The example

$$(6.3) \qquad \sum_{k=1}^{\infty} e^{-k} e^{-ks},$$

for which $\sigma_c = -1$, shows that Theorem 6.1 is false when $\alpha = 0$. If it is known in advance that σ_c is positive then the formula in hypothesis 1 will give it. For then α cannot be ≤ 0. If it were, we should have for any $\varepsilon > 0$ that

$$\sum_{k=1}^{n} a_k = O(\exp \lambda_n \varepsilon), \qquad n \to \infty.$$

This would imply, by Lemma 6.1, that (6.2) converges for $\sigma > \varepsilon$, a contradiction if $\varepsilon < \sigma_c$. But if $\alpha > 0$ Theorem 6.1 guarantees that $\sigma_c = \alpha$. It can be shown that if $\sigma_c < 0$, then

$$(6.4) \qquad \sigma_c = \varlimsup_{n \to \infty} \frac{\log |\sum_{k=n}^{\infty} a_k|}{\lambda_n}.$$

See Exercise 8 at the end of this chapter.

Corollary 6.1.

$$1. \quad \varlimsup_{n \to \infty} \frac{\log \sum_{k=1}^{n} |a_k|}{\lambda_n} = \alpha \qquad (or \ +\infty);$$

$$2. \quad \alpha > 0$$

$$\Rightarrow \qquad \sigma_a = \alpha \qquad (or \ +\infty).$$

The proof is immediate. It is equally apparent that if σ_a is known to be positive it is given by the formula of the corollary.

We can now show that the maximum difference between σ_a and σ_c is governed by the rate of increase of the exponents λ_k.

Theorem 6.2.

$$\sigma_a - \sigma_c \leq \varlimsup_{n \to \infty} \frac{\log n}{\lambda_n} = \beta,$$

To prove this we have only to show that if

$$(6.5) \qquad \sum_{k=1}^{\infty} a_k \exp(-\lambda_k s)$$

converges at $s = s_0$ then it converges absolutely at $s = s_0 + \beta\mu$ for any $\mu > 1$. If $\beta = +\infty$ there is nothing to prove. If $\beta < \infty$, the convergence of (6.5) at $s = s_0$ implies that

$$|a_k \exp(-\lambda_k \sigma_0)| < M, \qquad \text{some } M, \quad k = 1, 2, \ldots.$$

Thus

$$(6.6) \qquad \sum_{k=1}^{\infty} a_k \exp[-\lambda_k(s_0 + \beta\mu)] \ll M \sum_{k=1}^{\infty} \exp(-\lambda_k \beta\mu).$$

Choose a number p between 1 and μ. By definition of β

$$\frac{\log k}{\lambda_k} < \frac{\beta\mu}{p}$$

for all large k, since $\mu > p > 1$. But this implies that

$$\exp(-\lambda_k \beta\mu) < k^{-p},$$

so that series (6.6) is dominated by the convergent series

$$M \sum_{k=1}^{\infty} k^{-p},$$

and the proof is complete.

For ordinary Dirichlet series, $\lambda_n = \log n$, $\beta = 1$, and $\sigma_a - \sigma_c \leq 1$. That the difference may equal 1 is seen by the example

$$\eta(s) = \sum_{n=1}^{\infty} \frac{(-1)^{n+1}}{n^s},$$

for which $\sigma_a = 1$, $\sigma_c = 0$. For power series, $\beta = 0$, and we see again that in this case σ_a must equal σ_c.

7. Uniqueness

It is a simple matter to show that a given function $f(s)$ cannot be the sum of two different Dirichlet series. Note first that two series

$$(7.1) \qquad \sum_{k=1}^{\infty} a_k \exp(-\lambda_k s), \qquad \sum_{k=1}^{\infty} b_k \exp(-\mu_k s)$$

of apparently different types may be considered to be of the same type

$$(7.2) \qquad \sum_{k=1}^{\infty} c_k \exp(-v_k s).$$

One has only to define the sequence $\{v_k\}^{\infty}$ so as to include all of the λ_k and all of the μ_k of (7.1) and then to define the coefficients c_k so as to have (7.2) reduce to one or the other of the series (7.1). Thus to prove the uniqueness of representation it is clearly sufficient to prove the following result.

Theorem 7.1.

$$1. \quad \sum_{k=1}^{\infty} a_k \exp(-\lambda_k s) \equiv 0, \qquad \sigma > \sigma_c;$$

$$\Rightarrow \qquad a_k = 0, \qquad k = 1, 2, 3, \ldots.$$

For, suppose the contrary. Let a_j be the first coefficient not zero. Then

$$\sum_{k=j}^{\infty} a_k \exp[-(\lambda_k - \lambda_j)s]$$

is a Dirichlet series which vanishes for $\sigma > \sigma_c$. Allowing s to become infinite along the real axis we obtain by use of equation (5.5)

$$0 = \lim_{\sigma \to \infty} \sum_{k=j}^{\infty} a_k \exp[-(\lambda_k - \lambda_j)\sigma] = a_j.$$

The contradiction proves the theorem.

Let us establish two other results of somewhat the same character.

Theorem 7.2.

$$1. \quad f(s) = \sum_{k=1}^{\infty} a_k \exp(-\lambda_k s), \qquad \sigma > \sigma_c;$$

$$2. \quad a_1 \neq 0$$

$\Rightarrow \quad f(s)$ has at most a finite number of zeros in any

$$\mathrm{St}(s_0), \quad \sigma_0 > \sigma_c.$$

For, if $f(s)$ had infinitely many zeros in $\mathrm{St}(s_0)$ they would either cluster at a finite point or their real parts would tend to $+\infty$. In the former case $f(s)$ would be identically zero since $f(s)$ would be analytic at a cluster point of its zeros. But $f(s) \not\equiv 0$ by hypothesis 2, and Theorem 7.1. In the latter case the proof is completed as for Theorem 7.1 except that s is now allowed to become infinite through the zeros of $f(s)$ instead of along the real axis.

Note that the possibility of a cluster point of zeros on the axis of convergence is not ruled out by this theorem. Indeed this may occur. See Exercise 14 at the end of the chapter. Furthermore H. Bohr [1910] has shown that there may be infinitely many zeros in every right half-plane but not lying in any Stolz region. In contrast, we now show that if the above series converged absolutely $f(s)$ would be free of zeros in some right half-plane.

Theorem 7.3.

$$1. \quad f(s) = \sum_{k=1}^{\infty} a_k \exp(-\lambda_k s), \qquad \sigma > \sigma_a, \quad \sigma_a < \infty,$$

$$2. \quad a_1 \neq 0$$

$\Rightarrow \qquad\qquad f(s) \neq 0, \qquad \sigma > c, \quad \text{some } c.$

For

$$|f(s) - a_1 \exp(-\lambda_1 s)| \leqq \sum_{k=2}^{\infty} |a_k| \exp(-\lambda_k \sigma), \qquad \sigma > \sigma_a,$$

$$\sum_{k=2}^{\infty} |a_k| \exp(-\lambda_k b) \exp[-\lambda_k(\sigma - b)]$$

$$\leqq \exp[-\lambda_2(\sigma - b)] \sum_{k=2}^{\infty} |a_k| \exp(-\lambda_k b), \qquad \sigma > b > \sigma_a$$

$$|f(s)| \geqq |a_1| \exp(-\lambda_1 \sigma) - A \exp(-\lambda_2 \sigma),$$

$$A = \exp b\lambda_2 \sum_{k=2}^{\infty} |a_k| \exp(-\lambda_k b),$$

$$|f(s)| \geqq \exp(-\lambda_1 \sigma)[|a_1| - A \exp(-(\lambda_2 - \lambda_1)\sigma)].$$

Since the right-hand side becomes positive for large σ, the theorem is proved. This result shows, for example, that the example of Bohr must have $\sigma_a = \infty$.

8. Behavior on Vertical Lines

We can show quite trivially that if $f(\sigma + i\tau)$ is the sum of a Dirichlet series then its modulus cannot increase more rapidly than $|\tau|$ as $|\tau| \to \infty$.

Theorem 8.1.

$$1. \quad f(s) = \sum_{k=1}^{\infty} a_k \exp(-\lambda_k s), \qquad \sigma > \sigma_c;$$

$$2. \quad \sigma_1 > \sigma_c$$

$$\Rightarrow \qquad f(\sigma_1 + i\tau) = O(|\tau|), \qquad |\tau| \to \infty.$$

To prove this we apply Lemma 3 with $s_0 = \sigma_0$, $\sigma_c < \sigma_0 < \sigma_1$:

$$|f(\sigma_1 + i\tau)| \leqq M \frac{|\sigma_1 - \sigma_0 + i\tau|}{\sigma_1 - \sigma_0} \exp[-\lambda_1(\sigma_1 - \sigma_0)], \qquad \text{some } M$$

$$\leqq M\left(1 + \frac{|\tau|}{\sigma_1 - \sigma_0}\right) = O(|\tau|), \qquad |\tau| \to \infty.$$

We can improve Theorem 8.1 in two ways: replace $O(|\tau|)$ by $o(|\tau|)$ and make the result uniform in σ, $\sigma_1 \leqq \sigma < \infty$. Both improvements will be important for us in our inversion theory.

Theorem 8.2.

$$1. \quad f(s) = \sum_{k=1}^{\infty} a_k \exp(-\lambda_k s), \qquad \sigma > \sigma_c;$$

$$2. \quad \sigma_1 > \sigma_c$$

$$\Rightarrow \quad f(\sigma + i\tau) = o(|\tau|) \quad \text{uniformly in} \quad \sigma_1 \leqq \sigma < \infty \quad \text{as} \quad |\tau| \to \infty.$$

We wish to show that

$$\lim_{|\tau| \to \infty} \frac{f(\sigma + i\tau)}{|\tau|} = 0, \qquad \text{uniformly in } \sigma_1 \leqq \sigma < \infty.$$

Choose σ_0 as before and apply Lemma 3 to the truncated series

$$\left| \sum_{n}^{\infty} a_k \exp(-\lambda_k s) \right| \leqq M \, \frac{|\sigma - \sigma_0 + i\tau|}{\sigma - \sigma_0} \, \exp[-\lambda_n(\sigma - \sigma_0)],$$

$$\text{some } M, \quad \sigma > \sigma_0.$$

Then for $\sigma \geqq \sigma_1$

$$\frac{|f(\sigma + i\tau)|}{|\tau|} \leqq \frac{1}{|\tau|} \sum_{k=1}^{n-1} |a_k| \, \exp(-\sigma_1 \lambda_k)$$

$$+ M\left(\frac{1}{|\tau|} + \frac{1}{\sigma_1 - \sigma_0} \right) \exp[-\lambda_n(\sigma_1 - \sigma_0)].$$

Note that the right-hand side of this inequality is independent of σ in $\sigma_1 \leqq \sigma < \infty$. If we can show that it tends to zero as $|\tau| \to \infty$ our result will be established. But

$$(8.1) \qquad \varlimsup_{|\tau| \to \infty} \frac{|f(\sigma + i\tau)|}{|\tau|} \leqq \frac{M}{\sigma_1 - \sigma_0} \exp[-\lambda_n(\sigma_1 - \sigma_0)].$$

The left-hand side is independent of n, so that we may allow n to become ∞ in (8.1) to obtain

$$\varlimsup_{|\tau| \to \infty} \frac{|f(\sigma + i\tau)|}{|\tau|} = 0, \qquad \text{uniformly in } \sigma_1 \leqq \sigma < \infty,$$

from which the desired result follows.

For a deeper study of the behavior of f on vertical lines it is useful to define the order of increase as follows.

Definition 8. The order of increase of $f(\sigma + i\tau)$ on the line $\sigma = \sigma_1$ is

$$(8.2) \qquad \mu(\sigma_1) = \varlimsup_{|\tau| \to \infty} \frac{\log |f(\sigma_1 + i\tau)|}{\log |\tau|}.$$

It is easy to see that $\mu(\sigma_1)$ is the lower bound of numbers r such that

$$f(\sigma_1 + i\tau) = O(|\tau|^r), \qquad |\tau| \to \infty.$$

For example, if $f(s) = s^p$, then $\mu(\sigma) = p$ (p real), the order being the same on every vertical line. Theorem 8.2 shows that $\mu(\sigma) \leq 1$ for the function there defined. It is easy to see that $\mu(\sigma) = 0$ for $\sigma > \sigma_a$. Many properties of $\mu(\sigma)$ are known for a function defined as in Theorem 8.2. For example, it is continuous, nonnegative, non-increasing. Its actual determination is usually very difficult. It is of interest that $\mu(\sigma)$ may equal 1, showing that the conclusion of Theorem 8.2 is best possible in a certain sense.

9. Inversion

The inversion problem for a Dirichlet series expansion is the determination of the coefficients in terms of the expanded function. We develop here a formula that is analogous to the classical Cauchy inversion of a power series,

$$F(z) = \sum_{n=0}^{\infty} a_k z^k, \qquad a_n = \frac{1}{2\pi i} \int_\Gamma \frac{F(z)}{z^{n+1}} \, dz.$$

Our previous experience with Dirichlet series would indicate that we should seek rather the analog of a formula for the sum of the coefficients

$$(9.1) \qquad \frac{F(z)}{1-z} = \sum_{k=0}^{\infty} A_k z^k, \qquad A_n = \sum_{k=0}^{n} a_k.$$

$$(9.2) \qquad A_n = \frac{1}{2\pi i} \int_\Gamma \frac{F(z)}{(1-z)z^{n+1}} \, dz,$$

a formula which assumes that the circle Γ is in the region of convergence of series (9.1) and of radius less than one. After an exponential change of variable (9.2) becomes

$$A_n = \frac{1}{2\pi i} \int \frac{F(e^{-s})}{(1 - e^{-s})} e^{ns} \, ds,$$

the path of integration now being part of a vertical line. When the periodicity of $F(e^{-s})$ is abandoned, as it must be for the sum $f(s)$ of a general Dirichlet series, the factor $(1 - e^{-s})^{-1}$ may be replaced by s^{-1} (having the same residue at $s = 0$), and the path of integration may be taken as a whole vertical line. Without insisting on the details of this analogy we do wish to observe that it does show why this vertical line $\sigma = a$ must have $a > 0$ as well as $a > \sigma_c$.

We prove the inversion by use of two lemmas taken from the classical theory of residues.

Lemma 9.1.

$$1. \quad a > 0$$

$$(9.3) \quad \Rightarrow \qquad \frac{1}{2\pi i} \int_{a-i\infty}^{a+i\infty} \frac{e^{\omega s}}{s} \, ds = \begin{cases} 1, & \omega > 0, \\ 0, & \omega < 0. \end{cases}$$

We sketch the proof for the case $\omega > 0$. Integrating $e^{\omega s}/(2\pi i s)$ over the rectangle with vertices $b - iU, a - iU, a + iT, b + iT$ ($b < 0, T > 0, U > 0$) in the positive sense yields 1, the residue of the integrand at $s = 0$. For the integral over the left vertical side we have

$$\left| \int_{b-iU}^{b+iT} \frac{e^{\omega s}}{s} \, ds \right| \leqq \frac{e^{b\omega}}{(-b)} (T + U) = o(1), \qquad b \to -\infty.$$

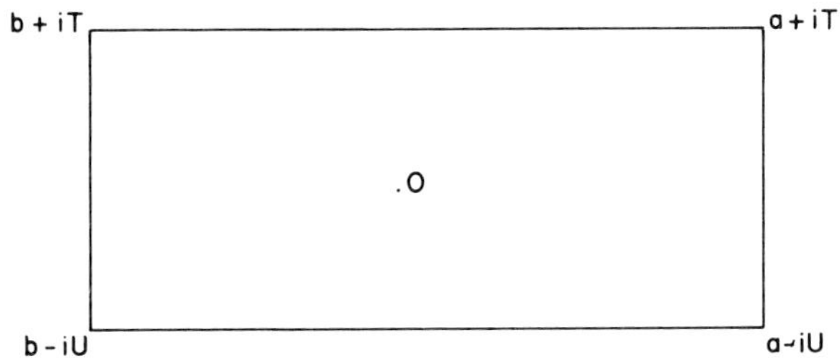

Figure 3

Hence

$$\frac{1}{2\pi i}\int_{a-iU}^{a+iT}\frac{e^{\omega s}}{s}\,ds - 1 = \frac{1}{2\pi i}\int_{-\infty}^{a}\frac{e^{\omega(\sigma+iT)}}{\sigma+iT}\,d\sigma - \frac{1}{2\pi i}\int_{-\infty}^{a}\frac{e^{\omega(\sigma-iU)}}{\sigma-iU}\,d\sigma$$

if either of the improper integrals converges. They both do so, as the following computation shows.

$$\left|\frac{1}{2\pi i}\int_{a-iU}^{a+iT}\frac{e^{\omega s}}{s}\,ds - 1\right| \leqq \frac{1}{2\pi}\int_{-\infty}^{a}\frac{e^{\omega\sigma}}{T}\,d\sigma + \frac{1}{2\pi}\int_{-\infty}^{a}\frac{e^{\omega\sigma}}{U}\,d\sigma$$

$$= \frac{e^{\omega a}}{2\pi\omega}\left(\frac{1}{T}+\frac{1}{U}\right).$$

Since the right-hand side tends to zero as T and U become infinite independently we see that the integral (9.3) converges and equals 1 for $\omega > 0$. The case $\omega < 0$ is handled similarly.

Lemma 9.2.

$$1. \quad a > 0$$

$$\Rightarrow \qquad \lim_{R\to\infty}\frac{1}{2\pi i}\int_{a-iR}^{a+iR}\frac{ds}{s} = \frac{1}{2}.$$

Since the integral is equal to

$$\frac{1}{2\pi i}\log\left(\frac{a+iR}{a-iR}\right) = \frac{1}{\pi}\tan^{-1}\frac{R}{a}$$

the result is immediate. Note that the integral (9.3) diverges when $\omega = 0$. But the divergent integral has the " Cauchy-value " 1/2, as shown by Lemma 9.2.

We now prove the inversion formula conjectured above.

Theorem 9.1.

$$1. \quad f(s) = \sum_{k=1}^{\infty} a_k \exp(-\lambda_k s), \qquad \sigma > \sigma_c;$$

$$2. \quad \lambda_n < \omega < \lambda_{u+1};$$

$$3. \quad a > \sigma_c, \qquad a > 0$$

(9.4) $\Rightarrow$

$$\sum_{k=1}^{n} a_k = \frac{1}{2\pi i}\int_{a-i\infty}^{a+i\infty}\frac{f(s)e^{\omega s}}{s}\,ds.$$

Set

$$g(s) = e^{\omega s} f(s) - \sum_{k=1}^{n} a_k \exp(\omega - \lambda_k)s = \sum_{k=n+1}^{\infty} a_k \exp[-(\lambda_k - \omega)s].$$

By Lemma 9.1

$$\frac{1}{2\pi i} \int_{a-i\infty}^{a+i\infty} \sum_{k=1}^{n} a_k \frac{\exp(\omega - \lambda_k)s}{s} \, ds = \sum_{k=1}^{n} a_k \, .$$

It remains to prove that

$$\int_{a-i\infty}^{a+i\infty} \frac{g(s)}{s} \, ds = 0.$$

Note that $g(s)$ is the sum of a Dirichlet series with first exponent positive. Hence if the series is divided by the exponential e^{-cs}, $c = \lambda_{n+1} - \omega$, it remains a Dirichlet series. Thus our theorem will be proved if we show that when $f(s)$ is defined as in hypothesis 1

$$(9.5) \qquad \int_{a-i\infty}^{a+i\infty} e^{-cs} \frac{f(s)}{s} \, ds = 0.$$

Integrating this integrand over the rectangle with vertices $a - iU$, $b - iU$, $b + iT$, $a + iT$ now yields zero since $g(s)/s$ is analytic inside the rectangle. But

$$\left| \int_{b-iU}^{b+iT} \frac{f(s)e^{-cs}}{s} \, ds \right| \leqq \int_{-U}^{T} \frac{|f(b + it)|}{b} e^{-cb} \, dt = o(1), \qquad b \to \infty$$

since $f(s)$ is uniformly bounded in the strip $-U \leqq \tau \leqq T$, $\sigma \geqq a$. See the remark following equation (3.6). Hence

$$(9.6) \qquad \left| \int_{a-iU}^{a+iT} \frac{f(s)e^{-cs}}{s} \, ds \right| \leqq \int_{a}^{\infty} \frac{|f(\sigma - iU)|}{U} e^{-c\sigma} \, d\sigma$$

$$+ \int_{a}^{\infty} \frac{|f(\sigma + iT)|}{T} e^{-c\sigma} \, d\sigma.$$

The improper integrals converge since $f(s)$ is bounded on the lines $\tau = -U$ and $\tau = T$. If $\varepsilon > 0$, then by Theorem 8.2 there exists a number R, independent of σ on (a, ∞), such that

$$(9.7) \quad |f(\sigma - iU)| < \varepsilon \, U, \qquad |f(\sigma + iT)| < \varepsilon \, T, \quad U > R, \quad T > R.$$

Applying (9.7) to (9.6), we obtain

$$\left| \int_{a-iU}^{a-iT} \frac{f(s)e^{-cs}}{s}\, ds \right| \leqq \frac{2\varepsilon}{c},$$

from which the convergence of the integral (9.5) to the value 0 is immediate.

Corollary 9.1.

$$1. \quad f(s) = \sum_{k=1}^{\infty} a_k \exp(-\lambda_k s), \qquad \sigma > \sigma_c;$$

$$2. \quad a > \sigma_c, \qquad a > 0$$

$$\Rightarrow \qquad \lim_{R \to \infty} \frac{1}{2\pi i} \int_{a-iR}^{a+iR} \frac{f(s)\exp \lambda_n s}{s}\, ds = \sum_{k=1}^{n} a_k - \frac{a_n}{2}.$$

For, by Theorem 9.1

$$\frac{1}{2\pi i} \int_{a-i\infty}^{a+i\infty} \frac{f(s) - a_n \exp(-\lambda_n s)}{s} \exp \lambda_n s \, ds = \sum_{k=1}^{n-1} a_k.$$

In order to break this convergent integral into two others, corresponding to the two terms in the bracket, we must use the Cauchy value of the resulting divergent integrals. Use of Lemma 9.2 then gives the desired result.

A notation which we shall use later for formula (9.4) is

$$\frac{1}{2\pi i} \int_{a-i\infty}^{a+i\infty} \frac{f(s)e^{\omega s}}{s}\, ds = \sum_{\lambda_k < \omega} a_k$$

for all ω different from the exponents λ_k. We shall later have occasion to use the following generalization of Theorem 9.1.

Theorem 9.2.

$$1. \quad f(s) = \sum_{k=1}^{\infty} a_k \exp(-\lambda_k s), \qquad \sigma > \sigma_c;$$

$$2. \quad \lambda_n \leqq \omega < \lambda_{n+1};$$

3. $a > \sigma_c, \quad a > 0, \quad p = 1, 2, 3, \dots$

$$\Rightarrow \qquad \frac{1}{2\pi i} \int_{a-i\infty}^{a+i\infty} \frac{f(s)e^{\omega s}}{s^{p+1}} \, ds = \sum_{k=1}^{n} \frac{a_k(\omega - \lambda_k)^p}{p!}.$$

Successive integration by parts gives

(9.8) $\qquad \dfrac{1}{2\pi i} \displaystyle\int_{a-i\infty}^{a+i\infty} \dfrac{f(s)e^{\omega s}}{s^{p+1}} \, ds = \dfrac{1}{2\pi i} \dfrac{1}{p!} \int_{a-i\infty}^{a+i\infty} \dfrac{[f(s)e^{\omega s}]^{(p)}}{s} \, ds.$

At each stage the integrated part vanishes. For example, at the final stage

$$\frac{[f(s)e^{\omega s}]^{(p-1)}}{s}$$

vanishes at $a + i\infty$ by Theorem 8.2, the numerator clearly being equal to $e^{\omega s}$ multiplied by the sum of a Dirichlet series. In fact for any positive integer p

(9.9) $\qquad [f(s)e^{\omega s}]^{(p)} = \displaystyle\sum_{k=1}^{\infty} a_k(\omega - \lambda_k)^p \exp(\omega - \lambda_k)s, \qquad \sigma > \sigma_c.$

Now apply Theorem 9.1 to the second integral (9.8) to obtain the desired result. Note that when $\omega = \lambda_n$ there is really no term for $k = n$ in (9.9), so we may still consider that ω is distinct from all existing exponents. We may thus write the value of (9.8) for *all* ω as

(9.10) $\qquad \dfrac{1}{p!} \displaystyle\sum_{\lambda_k \leqq \omega} a_k(\omega - \lambda_k)^p.$

For example, for the Riemann zeta function with $\lambda_k = \log k$ and $\omega = \log x$, we have

(9.11) $\qquad \displaystyle\sum_{k \leqq x} \log \frac{x}{k} = \frac{1}{2\pi i} \int_{a-i\infty}^{a+i\infty} \frac{\zeta(s)x^s}{s^2} \, ds, \qquad a > 1.$

10. A Mean-Value Theorem

We shall now show that the average value of $|f(s)|^2$ on a vertical line has a simple series expression involving the elements of the Dirichlet series for $f(s)$. We prove first a slightly more general result.

Theorem 10.1.

$$1. \quad f(s) = \sum_{k=1}^{\infty} a_k \exp(-\lambda_k s) \quad \text{converges absolutely for } \sigma = \alpha;$$

$$2. \quad g(s) = \sum_{k=1}^{\infty} b_k \exp(-\lambda_k s) \quad \text{converges absolutely for } \sigma = \beta.$$

$$(10.1) \quad \Rightarrow \quad \lim_{T \to \infty} \frac{1}{T} \int_0^T f(\alpha + iy)g(\beta - iy)\, dy$$

$$= \sum_{k=1}^{\infty} a_k b_k \exp[-\lambda_k(\alpha + \beta)],$$

the series (10.1) converging absolutely.

The double series

$$(10.2) \quad \sum_{k=1}^{\infty} \sum_{j=1}^{\infty} a_k b_j \exp[-\lambda_k(\alpha + iy) - \lambda_j(\beta - iy)]$$

converges (absolutely) since

$$(10.3) \quad \sum_{k=1}^{\infty} |a_k| \exp(-\lambda_k \alpha) \left[\sum_{j=1}^{\infty} |b_j| \exp(\lambda_j \beta) \right]$$

converges by hypothesis. The sum of (10.2) as summed by rows and columns is clearly the integrand of the integral (10.1). On the other hand if we isolate the terms in the principal diagonal ($j = k$) we obtain for the sum

$$(10.4) \quad \sum_{k=1}^{\infty} a_k b_k \exp[-\lambda_k(\alpha + \beta)]$$

$$+ \sum_{\substack{k=1 \\ k \neq j}}^{\infty} \sum_{j=1}^{\infty} a_k b_j \exp[-\lambda_k(\alpha + iy) - \lambda_j(\beta - iy)].$$

The double series (10.4), being dominated by the series of constants (10.3), converges uniformly for all y. Hence term-by-term integration yields

$$(10.5) \quad \frac{1}{T} \int_0^T f(\alpha + iy)g(\beta - iy)\, dy = \sum_{k=1}^{\infty} a_k b_k \exp[-\lambda_k(\alpha + \beta)]$$

$$+ \sum_{\substack{k=1 \\ k \neq j}}^{\infty} \sum_{j=1}^{\infty} a_k b_j \exp(-\lambda_k \alpha - \lambda_j \beta) \frac{1 - \exp iT(\lambda_j - \lambda_k)}{iT(\lambda_k - \lambda_j)}.$$

The double series (10.5) converges uniformly for all T since it is again dominated by a convergent series of constants. For,

$$\left| \frac{1 - e^{ix}}{ix} \right| \leqq 1, \qquad -\infty < x < \infty.$$

We now obtain our result by allowing T to become infinite in (10.5).

The mean-value theorem now follows as a corollary if we choose $b_n = \bar{a}_n$ and $\beta = \alpha$. Then $g(\beta - iy)$ is the conjugate of $f(\alpha + iy)$.

Corollary 10.1.

1. $f(s) = \sum_{k=1}^{\infty} a_k \exp(-\lambda_k s)$ converges absolutely for $\sigma = \alpha$

$$\Rightarrow \quad \lim_{T \to \infty} \frac{1}{2T} \int_{-T}^{T} |f(\alpha + iy)|^2 \, dy = \sum_{k=1}^{\infty} |a_k|^2 \exp(-2\lambda_k \alpha).$$

Note that we have here taken the average over the whole line $\sigma = \alpha$ rather than the half-line as in (10.1). Either result is clearly true.

We may use Theorem 10.1 to obtain a new inversion formula for Dirichlet series.

Theorem 10.2.

1. $f(s) = \sum_{k=1}^{\infty} a_k \exp(-\lambda_k s)$ converges absolutely for $\sigma = \alpha$;

2. $n = 1, 2, 3, \ldots$

$$\Rightarrow \qquad a_n = \lim_{T \to \infty} \frac{1}{T} \int_0^T \exp \lambda_n(\alpha + iy) f(\alpha + iy) \, dy.$$

This is a special case of Theorem 10.1 in which $b_k = 0$ for all $k \neq n$ and $b_n = 1$, $\beta = \alpha$.

11. Analytic Behavior of the Sum of a Dirichlet Series

Although a Dirichlet series is a generalization of a power series there is at least one marked difference in the theories of the two series. A function defined by a power series always has a singularity on the circle of convergence, but one defined by a Dirichlet series need have no singularity on the axis of convergence. The function (3.9) is a case in point for it is entire, as we shall prove in Chapter 3. See also Exercise 16 at the end of this chapter. However, if the coefficients of the series are all real and positive the sum of the series does have a singularity on the axis of convergence, as first proved by E. Landau [1905].

Theorem 11.1.

$$(11.1) \qquad 1. \quad f(s) = \sum_{k=1}^{\infty} a_k \exp(-\lambda_k s), \qquad \sigma > \sigma_c;$$

$$2. \quad a_k > 0, \qquad k = 1, 2, 3, \ldots$$

$$\Rightarrow \qquad f(s) \text{ is not analytic at } s = \sigma_c.$$

It is no restriction to assume $\sigma_c = 0$. For, under hypothesis 1

$$f(s + \sigma_c) = \sum_{k=1}^{\infty} [a_k \exp(-\lambda_k \sigma_c)] \exp(-\lambda_k s)$$

is a Dirichlet series with abscissa of convergence equal to zero. Its coefficients are also positive. Assume then that the origin is not a singularity of $f(s)$ in order to deduce a contradiction. In that case the Taylor expansion of $f(s)$ about $s = \delta > 0$,

$$f(s) = \sum_{k=0}^{\infty} f^{(k)}(\delta) \frac{(s - \delta)^k}{k!} = \sum_{k=0}^{\infty} \frac{(s - \delta)^k}{k!} \sum_{j=1}^{\infty} a_j (-\lambda_j)^k \exp(-\lambda_j \delta)$$

must converge at some point $s = -a < 0$. That is, the double series

$$\sum_{k=0}^{\infty} \sum_{j=1}^{\infty} \frac{(a + \delta)^k}{k!} a_j \lambda_j^k \exp(-\lambda_j \delta),$$

every term of which is positive, must converge. Hence we may sum in any order. But if we sum first with respect to k and then j we obtain

$$\sum_{j=1}^{\infty} a_j \exp(-\lambda_j \delta) \sum_{k=0}^{\infty} \frac{(a+\delta)^k}{k!} \lambda_j{}^k = \sum_{j=1}^{\infty} a_j \exp a\lambda_j.$$

This indicates that (11.1) converges at a point $s = -a$ to the left of the axis of convergence, an impossibility.

As an example, the zeta-function for which $a_k = 1$ must have a singularity at $s = 1$, a fact which we shall confirm by other methods in Chapter 3.

We show next that a function, not a constant, which is analytic at infinity cannot be the sum of a Dirichlet series. By contrast we shall show in Chapter 5 that such a function can be the generating function of a Laplace transform.

Theorem 11.2.

$$(11.2) \qquad 1. \quad \sum_{k=1}^{\infty} a_k \exp(-\lambda_k s) = \sum_{k=1}^{\infty} b_k s^{-k}, \qquad \sigma > R, \quad \text{some } R$$

$$\Rightarrow \qquad a_k = b_k = 0, \qquad k = 1, 2, \ldots.$$

Denote the sum of the Dirichlet series (11.2) by $f(s)$ and suppose that it is not identically zero. It is no restriction to suppose that $a_1 \neq 0$. Then $\lambda_1 > 0$ since the power series (11.2) vanishes at ∞. Denote the first nonzero b by b_m, so that

$$\lim_{\sigma \to +\infty} f(\sigma)\sigma^m = b_m \neq 0.$$

But by equation (5.6)

$$\lim_{\sigma \to +\infty} f(\sigma)\sigma^m = \lim_{\sigma \to +\infty} [f(\sigma) \exp \lambda_1\sigma][\exp(-\lambda_1\sigma)\sigma^m],$$

$$b_m = a_1 \lim[\exp(-\lambda_1\sigma)\sigma^m] = 0.$$

This is a clear contradiction, so that $f(s)$ must be identically zero. The desired conclusions now follow from Theorem 7.1 and the familiar uniqueness property for power series.

12. Summary

By way of summarizing the more important results of the present chapter let us list them briefly here in juxtaposition with the corresponding ones for power series.

$$F(z) = \sum_{k=0}^{\infty} a_k z^k \qquad\qquad f(s) = \sum_{k=1}^{\infty} a_k \exp(-\lambda_k s)$$

1.	Convergence	$|z| < p$	$\sigma > \sigma_c$
2.	Differentiation	$F'(z) = \sum_{k=0}^{\infty} k a_k z^{k-1}$	$f'(s) = -\sum_{k=1}^{\infty} a_k \lambda_k \exp(-\lambda_k s)$
3.	Analyticity	$|z| < p,$	$\sigma > \sigma_c$
4.	Uniqueness	$F \equiv 0 \Rightarrow a_k \equiv 0$	$f \equiv 0 \Rightarrow a_k \equiv 0$
5.	Inversion	$c < p, \quad c < 1$	$c > \sigma_c, \quad c > 0$

$$\sum_{k<n+1} a_k = \frac{1}{2\pi i} \int_{|z|=c} \frac{F(z)\,dz}{(1-z)z^{n+1}} \qquad\qquad \sum_{\lambda_k<\omega} a_k = \frac{1}{2\pi i} \int_{c-i\infty}^{c+i\infty} \frac{f(s)}{s} e^{s\omega}\,ds$$

6. $a_k > 0 \Rightarrow$ singularity at $z = p$ $\qquad\qquad$ $a_k > 0 \Rightarrow$ singularity at $s = \sigma_c$

We also list here two important points where analogy with power series fails. A function $f(s)$ defined by a Dirichlet series
(a) is never analytic at $s = \infty$ (unless it is a constant);
(b) need have no singularity on the axis of convergence.

EXERCISES

1. Find σ_a and σ_c for series (3.9) by use of Theorem 6.1 even though $\alpha = 0$. Make a preliminary translation in the s-plane.

2. Find σ_a and σ_c for an ordinary Dirichlet series with $a_k = (-1)^k/\sqrt{k}$.

3. Check the following table, thus showing all conceivable disparity between σ_a and σ_c.

a_k	λ_k	σ_c	σ_a
$\dfrac{1}{k!}$	$\log k$	$-\infty$	$-\infty$
$\dfrac{(-1)^k}{k}$	$\log \log k$	$-\infty$	1
$\dfrac{(-1)^k}{\sqrt{k}}$	$\log \log k$	$-\infty$	$+\infty$
$(-1)^k$	$\log k$	0	1
$(-1)^k$	$\log \log k$	0	$+\infty$
$k!$	$\log k$	$+\infty$	$+\infty$

4. Show that Lemma 6.1 is false if $\alpha < 0$. Take $\lambda_k = 2^k$ and

$$\sum_{k=1}^{n} a_k = \exp(-2^n).$$

Then $\alpha = -1$, but the corresponding Dirichlet series diverges at $s = -1/2$.

5. Prove Theorem 6.1 if $\alpha = +\infty$.

6. Prove that if

(1) $$\sum_{k=n}^{\infty} a_k = O(\exp \alpha \lambda_n), \qquad n \to \infty, \quad \alpha < 0$$

then

(2) $$\sum_{k=1}^{\infty} a_k \exp(-\lambda_k s)$$

converges for $\sigma > \alpha$.

7. Prove that if (2) of Exercise 6 converges for $s = \alpha < 0$ then (1) holds.

8. Prove formula (6.4).

9. Use (6.4) to find σ_a for an ordinary Dirichlet series with $a_k = e^{-k}$. Use this example to show Theorem 6.1 false if $\alpha = 0$.

10. Find σ_c for an ordinary Dirichlet series with $a_k = \dfrac{1}{[k(k+1)(k+2)]}$.

Ans. -2.

11. Same problem if $a_k = 1$ when k is a perfect cube, $a_k = 0$ otherwise.
Ans. $1/3$.

12. Use Theorem 5 to obtain Abel's classic continuity theorem for power series.

13. Prove Lemma 9.1 when $\omega < 0$.

14. From the general theory of analytic functions it is clear that $\sin(1 - e^{-s})^{-1}$ has a Dirichlet series expansion $(\lambda_n = n)$. Find σ_c and show that the sum of the series has many cluster points of its zeros on the axis of convergence.

15. Prove Theorem 9.2 if the integer p is replaced by an arbitrary positive number and $p!$ is replaced by $\Gamma(p+1)$.

16. If

$$\lambda_{2n} = n \qquad\qquad a_{2n} = 1$$

$$\lambda_{2n+1} = n + \frac{1}{n!} \qquad a_{2n+1} = -1,$$

show that $\sigma_c = 0$ for the corresponding Dirichlet series. Show that the sum of the series is an entire function. [Group terms two at a time to get a series that converges for all s.]

17. Prove that if $\lambda_n = n \log n$ then

$$\sigma_a = \varlimsup_{n \to \infty} \frac{\log |a_n|}{\lambda_n}.$$

18. Prove that

$$\frac{1}{2\pi i} \int_{2-i\infty}^{2+i\infty} \frac{\zeta(s)x^s}{s^{p+1}}\, ds = \sum_{k \le x} \left(\log \frac{x}{k}\right)^p, \qquad p = 2, 3, 4, \ldots.$$

19. Find the value of the following integral for every value of b for which it converges; find the Cauchy principal value when it diverges:

$$\int_{1-i\infty}^{1+i\infty} \frac{(2e^{-2s} + 2^{-s})e^{bs}}{s}\, ds.$$

20. Find the average value of $n^{2+iy}\zeta(2 + iy)$, $n = 1, 2, \ldots$, on the infinite interval $0 \le y < \infty$.   Ans. 1.

3 *The Zeta-Function*

1. Introduction

It is fitting that we should study $\zeta(s)$ in some detail. Its general properties will of course confirm and illustrate the results of Chapter 2, but in addition it will have specific properties resulting from its special definition. It is particularly these latter which we shall need in Chapter 4 in our proof of the prime number theorem. Moreover, this function deserves attention because of the great influence it has had on the development of mathematics, an influence undoubtedly resulting from the fact that it has thus far been impossible to locate exactly its zeros. To this extent the zeta-function, in spite of the simplicity of its definition, remains an enigma.

2. Analytic Nature of $\zeta(s)$

The zeta-function, defined by the Dirichlet series

$$\zeta(s) = \sum_{k=1}^{\infty} \frac{1}{k^s},$$

which converges for $\sigma > 1$, is analytic there. We now prove that it can be extended analytically, at least to $\sigma > 0$, and that it will be analytic there except for a single pole at $s = 1$.

Theorem 2. $\zeta(s)$ is analytic for $\sigma > 0$ except for a pole of order 1 and residue 1 at $s = 1$.

Denote by $[x]$ the largest integer $\leqq x$. Then by direct integration

$$(2.1) \qquad \zeta(s) = s \int_1^\infty \frac{[x]}{x^{s+1}} \, dx. \qquad \sigma > 1.$$

Note that one may also express $\zeta(s)$ by a Stieltjes* integral

$$(2.2) \qquad \zeta(s) = 1 + \int_1^\infty \frac{d[x]}{x^s}, \qquad \sigma > 1.$$

Then (2.1) follows from (2.2) by an integration by parts. We now accomplish analytic continuation into $\sigma > 0$ by adding and subtracting x to the numerator in (2.1):

$$(2.3) \qquad \zeta(s) = s \int_1^\infty \frac{[x] - x}{x^{s+1}} \, dx + \frac{s}{s-1}, \qquad \sigma > 1.$$

But now each term on the right has a meaning in the larger region $\sigma > 0$. If $\delta > 0$, the integral converges uniformly for $\sigma > \delta$,

$$(2.4) \qquad \int_1^\infty \frac{[x] - x}{x^{s+1}} \, dx \ll \int_1^\infty \frac{dx}{x^{\delta+1}} = \frac{1}{\delta}, \qquad \sigma \geqq \delta,$$

and hence represents an analytic function there. See also Exercise 12 at the end of this chapter. The rational term of (2.3) has a pole of residue 1 at $s = 1$, so that the proof is complete.

We give a second proof of the theorem, in some respects more elementary. Direct multiplication gives

$$(2.5) \qquad (1 - 2^{1-s})\zeta(s) = \eta(s) = \sum_{k=1}^\infty \frac{(-1)^{k+1}}{k^s}, \qquad \sigma > 1,$$

$$(1 - 3^{1-s})\zeta(s) = \xi(s) = 1 + \frac{1}{2^s} - \frac{2}{3^s} + \frac{1}{4^s} + \frac{1}{5^s} - \frac{2}{6^s} + \cdots,$$

$$\sigma > 1.$$

* For the definition of the Stieltjes integral and the simple properties thereof see, for example, D. V. Widder (1961, p. 149).

But these two Dirichlet series converge for $\sigma > 0$ since the partial sums of their coefficients are bounded (taking on only the values 1, 2, 0). They thus provide analytic continuations for the functions on the left side of the equations. That is,

$$\zeta(s) = \frac{\eta(s)}{1 - 2^{1-s}} = \frac{\xi(s)}{1 - 3^{1-s}}, \qquad \sigma > 0,$$

except perhaps where the denominators vanish:

$$s = 1 + \frac{2k\pi i}{\log 2}, \qquad s = 1 + \frac{2n\pi i}{\log 3}, \qquad k,n = 0, \pm 1, \pm 2, \ldots.$$

Since $s = 1$ is the only common point in these two sets of possible poles of $\zeta(s)$ (by the unique factorization theorem of elementary number theory), we see that $s = 1$ is the only pole for $\sigma > 0$. The residue there is

$$\lim_{s \to 1} \zeta(s)(s - 1) = \lim_{s \to 1} \eta(s) \frac{(s - 1)}{1 - 2^{1-s}} = \frac{\eta(1)}{\log 2} = 1.$$

This completes the proof. Incidentally we have found a set of zeros for $\eta(s)$ and another set for $\xi(s)$.

3. Euler Product for $\zeta(s)$

A very important property of $\zeta(s)$, defined by a series involving all positive integers, is that it can also be expressed as an infinite product involving all the primes. Denote the kth prime, arranged in increasing order, by p_k:

$$p_1 = 2, \quad p_2 = 3, \quad p_3 = 5, \quad p_4 = 7, \quad p_5 = 11, \ldots.$$

Then Euler's product development is given in the following theorem.

Theorem 3.

(3.1) $$\zeta(s) = \prod_{k=1}^{\infty} \frac{1}{1 - p_k^{-s}} \qquad \sigma > 1,$$

the product converging absolutely.

Note first that if one multiplies together the two series

$$\frac{1}{1-2^{-s}} = \sum_{n=0}^{\infty} \frac{1}{2^{ns}}, \qquad \frac{1}{1-3^{-s}} = \sum_{n=0}^{\infty} \frac{1}{3^{ns}},$$

absolutely convergent for $\sigma > 0$, the result is a series

$$\sum_{k=1}^{4} \frac{1}{k^s} + \frac{1}{6^s} + \frac{1}{8^s} + \frac{1}{9^s} + \frac{1}{12^s} + \frac{1}{16^s} + \cdots,$$

where the integers involved have only powers of 2 and 3 as factors. All coefficients are unity by the unique factorization theorem. In like manner

$$(3.2) \qquad \prod_{k=1}^{N} \frac{1}{1-p_k^{-s}} = \sum_{k=1}^{N} \frac{1}{k^s} + {\sum_{k>N}}' \frac{1}{k^s},$$

where the prime indicates that k runs through those integers, and only those, which are factorable in terms of the first N primes. The series (3.2) converges absolutely for $\sigma > 0$ since it was obtained as a product of such series. For $\sigma > 1$ we have

$$\left| \prod_{k=1}^{N} \frac{1}{1-p_k^{-s}} - \sum_{k=1}^{N} \frac{1}{k^s} \right| = {\sum_{k>N}}' \frac{1}{k^\sigma} \leqq \sum_{k=N+1}^{\infty} \frac{1}{k^\sigma}.$$

We have here strengthened our inequality by dropping the prime, but the resulting series no longer converges for $\sigma > 0$. We now see that for $\sigma > 1$ the right-hand side tends to zero as $N \to \infty$, and since the series on the left is known to approach $\zeta(s)$ as $N \to \infty$ the same must be true of the product.

By a familiar criterion the product (3.1) converges absolutely if

$$\sum_{k=1}^{\infty} \frac{1}{|p_k^s|} < \infty.$$

But this is only a part of the series of positive terms $\sum 1/k^\sigma$, known to converge for $\sigma > 1$. Hence it certainly converges there.

This result is basic in the study of the distribution of primes since it establishes a relation between all the positive integers on the one hand and all the primes on the other.

4. The Zeros of $\zeta(s)$

Euler's product shows that $\zeta(s) \neq 0$ when $\sigma > 1$. It of course gives no information in the further strip $0 < \sigma \leq 1$ to which we have extended the function analytically in §2. Indeed the exact distributions of the zeros there is still unknown. The famed Riemann hypothesis states that they all lie on the line $\sigma = \frac{1}{2}$, but this remains unproved. We prove only that there are no zeros on the right-hand boundary of this strip.

Theorem 4.

$$\zeta(1 + iy) \neq 0, \qquad -\infty < y < \infty.$$

We first call attention to the trigonometric identity and inequality

(4.1) $$2(1 + \cos \theta)^2 = 3 + 4 \cos \theta + \cos 2\theta \geq 0.$$

From Euler's product

$$\log \zeta(s) = -\sum_{k=1}^{\infty} \log (1 - p_k^{-s}), \qquad \sigma > 1,$$

where we use that branch of the logarithm which reduces to 0 at 1. Then

$$\log \zeta(s) = \sum_{k=1}^{\infty} \sum_{n=1}^{\infty} \frac{p_k^{-ns}}{n}, \qquad \sigma > 1.$$

Consequently

$$\log |\zeta(\sigma + iy)| = \sum_{k=1}^{\infty} \sum_{n=1}^{\infty} \frac{p_k^{-n\sigma}}{n} \cos (ny \log p_k).$$

Now using (4.1) we obtain

$$\log |\zeta^3(\sigma)\zeta^4(\sigma + iy)\zeta(\sigma + 2iy)|$$
$$= \sum_{k=1}^{\infty} \sum_{n=1}^{\infty} \frac{p_k^{-n\sigma}}{n} [3 + 4 \cos \theta + \cos 2\theta] \geq 0,$$

where $\theta = ny \log p_k$. Hence

(4.2) $$\zeta^3(\sigma) |\zeta(\sigma + iy)|^4 |\zeta(\sigma + 2iy)| \geq 1, \qquad \sigma > 1.$$

In order to deduce a contradiction, suppose that $\zeta(1 + iy_0) = 0$, $y_0 \neq 0$.

By Theorem 2 $\zeta(s)$ is analytic at $s = 1 + iy_0$ so that by Taylor's expansion

$$\zeta(\sigma + iy_0)^4 = O[(\sigma - 1]^4), \qquad \sigma \to 1-.$$

$\zeta(s)$ is also analytic at $s = 1 + 2iy_0$, so that whether it has a zero there or not

$$\zeta(\sigma + 2iy_0) = O(1), \qquad \sigma \to 1-.$$

Finally, by Theorem 2

$$\zeta^3(\sigma) = O([\sigma - 1]^{-3}), \qquad \sigma \to 1-.$$

Hence

$$\zeta^3(\sigma)|\zeta(\sigma + iy_0)|^4|\zeta(\sigma + 2iy_0)| = O(\sigma - 1) = o(1), \qquad \sigma \to 1-.$$

This contradicts inequality (4.2) and completes the proof.

5. Order of $\zeta(s)$ and $\zeta'(s)$ on Vertical Lines

Since $\sigma_a = 1$ for $\zeta(s)$ we know that $\zeta(s)$ must be bounded in any half-plane $\sigma \geq \sigma_1 > 1$. But general theory gives us no information about the behavior of $|\zeta(1 + iy)|$ as $|y| \to \infty$. We shall show here that it cannot increase more rapidly than $\log |y|$ and indeed that this behavior holds uniformly in $1 \leq \sigma \leq 2$. This uniformity will be of basic importance in our study of the distribution of primes.

Theorem 5.1.

$$\zeta(\sigma + iy) = O(\log |y|), \qquad |y| \to \infty$$

uniformly in $1 \leq \sigma \leq 2$.

From equation (2.2) we have

$$(5.1) \qquad \zeta(s) = \left(1 + \int_1^y \frac{d[x]}{x^s}\right) + \int_y^\infty \frac{d[x]}{x^s} = I_1 + I_2, \qquad \sigma > 1.$$

Note first that

$$|\zeta(\sigma - iy)| = |\zeta(\sigma + iy)|, \text{ so that it is no restriction to assume } y \geq 1.$$

Since $[x] \in \uparrow$ it is clear for $1 < \sigma \leq 2$ that

$$(5.2) \qquad |I_1| \leq 1 + \int_1^y \frac{d[x]}{x} = \frac{[y]}{y} + \int_1^y \frac{[x]}{x^2}\, dx \leq 1 + \log y.$$

Consider next the integral

$$J(s) = \int_y^\infty \frac{[x]}{x^{s+1}}\, dx = \int_y^\infty \frac{[x] - x}{x^{s+1}}\, dx + \int_y^\infty \frac{dx}{x^s}, \qquad \sigma > 1,$$

and note that

$$\left| J(s) - \frac{1}{(s-1)y^{s-1}} \right| \leq \int_y^\infty \frac{1}{x^2}\, dx = \frac{1}{y}, \qquad \sigma > 1$$

$$|J(s)| \leq \frac{1}{|s-1|} + \frac{1}{y} \leq \frac{2}{y}, \qquad 1 < \sigma \leq 2.$$

Integration by parts shows that

$$I_2 = \frac{-[y]}{y^s} + sJ(s), \qquad \sigma > 1$$

$$(5.3) \qquad |I_2| \leq 1 + (2 + y)\frac{2}{y} = O(1), \qquad 1 < \sigma \leq 2.$$

From (5.1), (5.2), and (5.3) it is clear that

$$(5.4) \qquad \qquad \zeta(\sigma + iy) = O(\log y), \qquad y \to +\infty$$

uniformly in $1 < \sigma \leq 2$. In the above calculations it was important that $\sigma > 1$. But now by Theorem 2 we know that $\zeta(\sigma + iy)$ is continuous for $\sigma \geq 1$, $y \geq 1$ so that (4) must also hold uniformly in $1 \leq \sigma \leq 2$.
We now prove a similar result for $\zeta'(s)$.

Theorem 5.2.

$$\zeta'(\sigma + iy) = O(\log^2 |y|), \qquad |y| \to \infty$$

uniformly in $1 \leq \sigma \leq 2$.

By Theorem 4 of Chapter 2 we have

$$(5.5) \qquad -\zeta'(s) = \int_1^y \frac{\log x}{x^s}\, d[x] + \int_y^\infty \frac{\log x}{x^s}\, d[x] = I_1 + I_2, \qquad \sigma > 1,$$

$$(5.6) \qquad |I_1| \leqq \log y \int_1^y \frac{d[x]}{x} \leqq \log y + \log^2 y, \qquad \sigma > 1.$$

Consider next the integral

$$(5.7) \qquad K(s) = \int_y^\infty \frac{[x] \log x}{x^{s+1}} \, dx = \int_y^\infty \frac{([x] - x)}{x^{s+1}} \log x \, dx + \int_y^\infty \frac{\log x}{x^s} \, dx.$$

The latter integral can be computed:

$$(5.8) \qquad \int_y^\infty \frac{\log x}{x^s} \, dx = \frac{\log y}{(s-1)y^{s-1}} + \frac{1}{(s-1)^2 y^{s-1}}, \qquad \sigma > 1,$$

$$(5.9) \qquad \left| \int_y^\infty \frac{\log x}{x^s} \, dx \right| \leqq \frac{\log y}{y} + \frac{1}{y^2} \qquad \sigma > 1, \quad y \geqq 1.$$

Hence from (5.7), (5.8), and (5.9) we have

$$|K(s)| \leqq \frac{\log y}{y} + \frac{1}{y^2} + \int_y^\infty \frac{\log x}{x^2} \, dx = \frac{2 \log y}{y} + \frac{1}{y^2} + \frac{1}{y}.$$

By an integration by parts we can now express I_2 in terms of $K(s)$ and $J(s)$:

$$I_2 = \frac{-[y] \log y}{y^s} + sK(s) - J(s), \qquad \sigma > 1,$$

$$(5.10) \qquad |I_2| \leqq \log y + (2 + y)|K(s)| + \frac{2}{y} = O(\log y), \qquad y \to \infty.$$

The dominant term is $\log^2 y$, resulting from I_1. The proof is now completed as for Theorem 5.1.

6. The Reciprocal of $\zeta(s)$

We now obtain a result for $1/\zeta(s)$ similar to those of §5. We shall see that the rate of increase of $1/|\zeta(1 + iy)|$ is no greater than $\log^7 |y|$. A smaller power could be used, but we shall make no attempt to improve the result, since any positive power will be sufficient for our purposes in Chapter 4.

Theorem 6.

$$\left|\frac{1}{\zeta(\sigma + iy)}\right| = O(\log^7 |y|), \qquad |y| \to \infty$$

uniformly in $1 \leq \sigma \leq 2$.

From equation (4.2) we have

$$|\zeta(\sigma + iy)| \geq \frac{1}{\zeta^{3/4}(\sigma)|\zeta(\sigma + 2iy)|^{1/4}}, \qquad \sigma > 1.$$

Since $\zeta(s)$ has a simple pole at $s = 1$ and since $\zeta(\sigma + 2iy) = O(\log y)$ by Theorem 5.1 we obtain for some positive constant A

$$(6.1) \qquad |\zeta(\sigma + iy)| \geq \frac{A(\sigma - 1)^{3/4}}{\log^{1/4} y}, \qquad 1 < \sigma \leq 2, \quad y > 1.$$

This inequality alone cannot yield the desired uniformity since $\sigma - 1$ is not bounded away from zero in $1 \leq \sigma \leq 2$. We divide the infinite rectangle $1 \leq \sigma \leq 2$, $1 \leq y \leq \infty$ into two parts by a curve

$$\eta - 1 = C \log^{-9} y, \qquad \text{some } C > 0,$$

traced out by the point (η, y). In the upper part inequality (6.1) will be adequate. For in that region $\sigma \geq \eta$ so that (6.1) becomes

$$|\zeta(\sigma + iy)| \geq AC \log^{-7} y.$$

For the lower region where $\sigma < \eta$ we have

$$(6.2) \quad |\zeta(\sigma + iy) - \zeta(\eta + iy)| \leq \int_\sigma^\eta |\zeta'(x + iy)| \, dx \leq B(\eta - 1) \log^2 y,$$

$$\text{some } B > 0,$$

by Theorem 5.2. Hence from (6.1), with σ replaced by η, and (6.2)

$$|\zeta(\sigma + iy)| \geq |\zeta(\eta + iy)| - B(\eta - 1) \log^2 y$$

$$\geq \frac{A(\eta - 1)^{3/4}}{\log^{1/4} y} - B(\eta - 1) \log^2 y = (AC^{3/4} - BC) \log^{-7} y.$$

If we now choose C so that $AC^{3/4} - BC > 0$ we see that $|\zeta(\sigma + iy)| \log^7 y$ is bounded away from zero throughout the rectangle $1 < \sigma \leq 2$, $1 \leq y < \infty$. Since $\zeta(s)$ has no zeros on the line $\sigma = 1$ its reciprocal is continuous in the strip $1 \leq \sigma \leq 2$ and our theorem is proved.

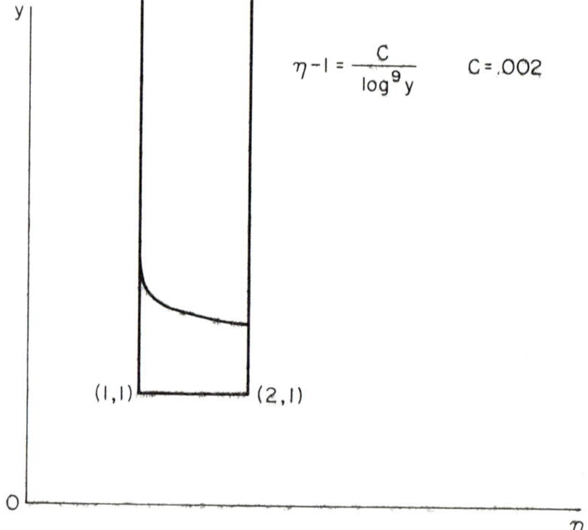

Figure 1

7. The Functional Equation for $\zeta(s)$

The zeta-function was defined by a Dirichlet series convergent only for $\sigma > 1$, but in §2 its definition was extended to the half-plane $\sigma > 0$. We here complete the definition for all s and derive the functional equation of Riemann which the completely defined function satisfies. We shall need a few preliminary results.

Lemma 7.1.

$$1. \quad a > 0, \ -1 < \sigma < 1$$

$$(7.1) \quad \Rightarrow \quad \int_0^\infty \frac{x^\sigma}{x^2 + a^2}\, dx = \frac{\pi}{2} a^{\sigma - 1} \sec \frac{\pi\sigma}{2}.$$

By the change of variable $x^2 = a^2 t$ integral (7.1) becomes

$$\frac{a^{\sigma - 1}}{2} \int_0^\infty \frac{t^{(\sigma - 1)/2}}{t + 1}\, dt = \frac{a^{\sigma - 1}}{2} \Gamma\!\left(\frac{1 + \sigma}{2}\right)\Gamma\!\left(\frac{1 - \sigma}{2}\right).$$

Now the result is established by use of the familiar identity

$$(7.2) \quad \Gamma(\sigma)\Gamma(1 - \sigma) = \frac{\pi}{\sin \pi\sigma}, \quad 0 < \sigma < 1.$$

Lemma 7.2.

$$1. \quad \sigma > 1$$

$$\Rightarrow \qquad \int_0^\infty \frac{x^{\sigma-1}}{e^x - 1}\, dx = \zeta(\sigma)\Gamma(\sigma).$$

This is proved by expanding the integrand in a series of powers of e^{-x} and integrating term by term:

$$(7.3) \qquad \int_0^\infty \frac{x^{\sigma-1}}{e^x - 1}\, dx = \sum_{k=1}^\infty \int_0^\infty x^{\sigma-1}\, e^{-kx}\, dx = \sum_{k=1}^\infty \frac{\Gamma(\sigma)}{k^\sigma} = \Gamma(\sigma)\zeta(\sigma).$$

The term by term integration is valid by the convergence theorem for monotone sequences, applicable here since the terms of the series are all positive and the series (7.3) for $\zeta(\sigma)$ is known to converge for $\sigma > 1$.

Lemma 7.3.

$$1. \quad x \neq 2k\pi i, \qquad k = 0, \pm 1, \pm 2, \ldots$$

$$\Rightarrow \qquad \frac{1}{e^x - 1} = \frac{1}{x} - \frac{1}{2} + \sum_{k=1}^\infty \frac{2x}{x^2 + 4k^2\pi^2}.$$

This is the Mittag–Leffler development of the meromorphic function on the left. We assume it known. See K. Knopp [1928, p. 419].

We now state Riemann's functional equation. It will be without meaning until $\zeta(s)$ is defined for $\sigma < 0$. This will be done in the course of the proof.

Theorem 7.1.

$$(7.4) \qquad \zeta(s) = 2^s \pi^{s-1} \sin\frac{\pi s}{2} \Gamma(1-s)\zeta(1-s), \qquad s \neq 1.$$

The desired analytic continuation will be accomplished by the simple observation that the rational function $1/s$ has two different integral representations in the right and left half-planes:

$$(7.5) \qquad \frac{1}{s} = \int_0^1 x^{s-1}\, dx \quad (\sigma > 0); \qquad \frac{1}{s} = -\int_1^\infty x^{s-1}\, dx \qquad (\sigma < 0).$$

By Lemma 7.2 we have for $\sigma > 1$ that

$$(7.6) \quad \Gamma(\sigma)\zeta(\sigma) = \int_0^1 \left[\frac{1}{e^x - 1} - \frac{1}{x}\right] x^{\sigma-1}\, dx + \frac{1}{\sigma - 1} + \int_1^\infty \frac{x^{\sigma-1}}{e^x - 1}\, dx.$$

Here we have used the first integral (7.5) replacing s by $\sigma - 1$. Since the bracket tends to $-\frac{1}{2}$ as $x \to 0+$ the first integral (7.6) converges for $\sigma > 0$ and consequently is the restriction to the real axis of a function analytic in the half-plane $\sigma > 0$. Similarly the last integral (7.6) defines an entire function of s when σ is replaced by s therein, and the right-hand side of (7.6) will be analytic for $\sigma > 0$ except for a pole at $s = 1$. Equation (7.6) shows that it coincides with $\Gamma(s)\zeta(s)$ on part of the real axis. We have thus extended $\zeta(s)$ analytically anew into the larger half-plane $\sigma > 0$. Using the second integral (7.5) with s replaced by $\sigma - 1$, (7.6) becomes

$$\Gamma(\sigma)\zeta(\sigma) = \int_0^\infty \left[\frac{1}{e^x - 1} - \frac{1}{x}\right] x^{\sigma-1}\, dx, \qquad 0 < \sigma < 1.$$

We now repeat this process, addding and subtracting the first integral (7.5) multiplied by $(\frac{1}{2})$:

$$(7.7) \quad \Gamma(\sigma)\zeta(\sigma) = \int_0^1 \left[\frac{1}{e^x - 1} - \frac{1}{x} + \frac{1}{2}\right] x^{\sigma-1}\, dx$$

$$- \frac{1}{2\sigma} + \int_1^\infty \left[\frac{1}{e^x - 1} - \frac{1}{x}\right] x^{\sigma-1} dx:$$

The bracket of the first integral is $O(x)$ as $x \to 0+$. Hence that integral converges for $\sigma > -1$. We are thus provided with an analytic continuation of $\zeta(s)$ into the half-plane $\sigma > -1$. Using the second integral (7.5) we obtain

$$(7.8) \quad \Gamma(\sigma)\zeta(\sigma) = \int_0^\infty \left[\frac{1}{e^x - 1} - \frac{1}{x} + \frac{1}{2}\right] x^{\sigma-1}\, dx, \qquad -1 < \sigma < 0.$$

From Lemmas 7.1 and 7.3

$$(7.9) \quad \Gamma(\sigma)\zeta(\sigma) = \sum_{k=1}^\infty \int_0^\infty \frac{2x^\sigma}{x^2 + 4\pi^2 k^2}\, dx = \frac{\pi}{\cos \dfrac{\pi\sigma}{2}} \sum_{k=1}^\infty (2k\pi)^{\sigma-1}.$$

This operation is valid since the terms of the series are all positive and since the resulting series on the right is known to converge for $-1 < \sigma < 0$. Thus the function $\zeta(s)$, newly defined by virtue of (7.8), is found to be expressed by (7.9) in terms of the original defining Dirichlet series. But now the equation (7.9) itself may be used for completing the definition of $\zeta(s)$ for $\sigma < -1$, since the series on the right converges there. That (7.9) is equivalent to the functional equation as stated above may be seen if we replace σ by $(1 - s)$ therein.

We may use the functional equation to read off some of the basic properties of $\zeta(s)$. Since all of the factors on the right-hand side of equation (7.4) are known to be analytic for $\sigma < 0$, the same is true of $\zeta(s)$. Moreover it must have the same zeros as $\sin \pi s/2$, there: $s = -2$, $-4, -6, \ldots$. These are called the *trivial zeros* of $\zeta(s)$. By Theorem 2

$$\zeta(1 - s) \sim \frac{-1}{s}, \qquad s \to 1,$$

so that $\zeta(0) = -\frac{1}{2}$. From §2, above, or from Exercise 2 at the end of this chapter, it is clear that $\eta(\frac{1}{2}) > 0$ and hence that $\zeta(\frac{1}{2}) < 0$. By use of this fact and the known equation $\Gamma(\frac{1}{2}) = \sqrt{\pi}$ we may check the functional equation for $s = \frac{1}{2}$. But note that the equation gives no information about $\zeta(s)$ on the line $\sigma = \frac{1}{2}$. From equation (2.5) we see that $\eta(s)$ is an entire function, since the only singularity of $\zeta(s)$ is cancelled by the zero at $s = 1$ of the factor $(1 - 2^{1-s})$. We summarize these results in the following theorem.

Theorem 7.2. The function $\zeta(s)$ as defined for $\sigma > 1$ by

$$\zeta(s) = \sum_{k=1}^{\infty} \frac{1}{k^s}$$

is continuable as a meromorphic function over the entire s-plane with a unique pole of order 1 and residue 1 at $s = 1$. Its only real zeros are at the points $s = -2, -4, -6, \ldots$. The function $\eta(s)$ defined for $\sigma > 0$ by

$$\eta(s) = \sum_{k=1}^{\infty} \frac{(-1)^{k+1}}{k^s}$$

is continuable over the whole plane so as to be an entire function.

We conclude this chapter by recounting a few further known results about the zeros of $\zeta(s)$ and about its order on vertical lines. L. J. Comrie [1936] has computed the first 1041 of them, in the order of increase of their positive imaginary parts, and has found them all to lie on the line $\sigma = \frac{1}{2}$ and with imaginary parts < 1468, in accord with the Riemann hypothesis. The first six have approximate imaginary parts; J. P. Gram [1903, p. 297]:

$$14.13, \quad 21.02, \quad 25.01, \quad 30,42, \quad 32.93, \quad 37.58.$$

Let us denote by $N(T)$ the number of zeros of $\zeta(s)$ in the rectangle $0 \leq \tau \leq T$, $0 \leq \sigma \leq 1$ and by $N_0(T)$ the number of these on the line $\sigma = \frac{1}{2}$. If the Riemann hypothesis is true $N(T) = N_0(T)$ for all T. The following facts have been proved, by the authors indicated.

1. $N(\infty) = \infty$ B. Riemann [1859, p. 671]
2. $N_0(\infty) = \infty,$ G. H. Hardy [1914, p. 1012]

3. $N(T) \sim \dfrac{T \log T}{2\pi},$ $T \to \infty,$ B. Riemann [1859, p. 671]

4. $N_0(T) > AT \log T,$ some $A > 0.$ A. Selberg [1942, p. 101]

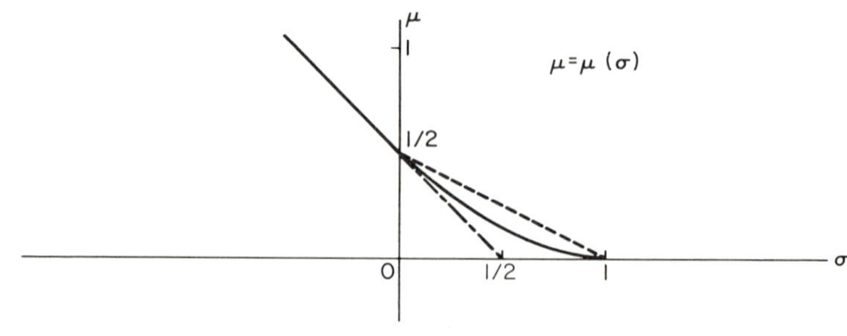

Figure 2

The graph of the function $\mu(\sigma)$ defined generally by (8.2) of Chapter 2 is of particular interest for the zeta-function. It is clear that the sum of any Dirichlet series is bounded on any vertical line inside its half-plane of absolute convergence. For,

$$|f(\sigma + iy)| \leq \sum_{k=1}^{\infty} |a_k| \exp(-\lambda_k \sigma) < \infty, \qquad \sigma > \sigma_a,$$

and the series on the right is independent of y. Hence $\mu(\sigma) = 0$ when $\sigma > \sigma_a$. For $\zeta(s)$ we know that $\sigma_a = 1$. Equation (7.4) enables us to compute $\mu(\sigma)$ also when $\sigma < 0$. For,

$$\left| \sin \frac{\pi}{2}(\sigma + iy) \right| \sim e^{\pi|y|/2}, \qquad |y| \to \infty,$$

$$|\Gamma(1 - \sigma - iy)| \sim (2\pi)^{1/2} |y|^{-\sigma + (1/2)} e^{-\pi|y|/2}, \qquad |y| \to \infty,$$

(see E. C. Titchmarsh [1939, p. 259]). The final factor on the right of (7.4) is bounded as $|y| \to \infty$ for any negative σ. That is,

$$|\zeta(\sigma + iy)| \sim A|y|^{-\sigma + (1/2)}, \qquad \text{some } A, \quad |y| \to \infty.$$

Hence

$$(7.10) \qquad \mu(\sigma) = \frac{1}{2} - \sigma, \qquad \sigma < 0.$$

But the exact nature of the graph of $\mu(\sigma)$ is unknown in the critical interval, $0 < \sigma < 1$. It is known to be continuous, nonincreasing, and concave upward for any Dirichlet series. See G. H. Hardy and M. Riesz [1915, p. 16]. One natural guess is that equation (7.10) holds for $\sigma < \frac{1}{2}$ and that $\mu(\sigma) = 0$ for $\sigma \geq \frac{1}{2}$, and this conjecture is known as the Lindelöf hypothesis. It is known to be a consequence of the Riemann hypothesis, but not conversely. Three possible variants of the graph are indicated in Fig. 2, the lowest of which is Lindelöf's [1908].

8. Summary

In this chapter we have extended the definition of $\zeta(s)$ and $\eta(s)$ to the whole plane, discovering that the first is meromorphic with a single pole, that the second is entire. We have expanded $\zeta(s)$ in Euler's infinite product, exhibiting the primes, and have obtained Riemann's basic functional equation for it. We have proved that $\zeta(1 + iy) \neq 0$ and that

$$(8.1) \qquad \frac{\zeta'(s)}{\zeta(s)} \sim -\frac{1}{s-1}, \qquad s \to 1,$$

$$(8.2) \qquad \frac{\zeta'(\sigma + iy)}{\zeta(\sigma + iy)} = O(\log^9 |y|), \qquad |y| \to \infty$$

uniformly in $1 \leq \sigma \leq 2$. We have indeed proved more than (8.1) and (8.2), but these are the facts as we shall need them in Chapter 4.

EXERCISES

1. Find the Dirichlet series for $(1 - 5^{1-s})\zeta(s)$ and find its absicissa of convergence.

2. If $\eta(s)$ is defined by (2.5) prove that

$$1 - 2^{-\sigma} < \eta(\sigma) < 1, \qquad \sigma > 0.$$

3. Give the details of deriving Theorem 7.1 from equation (7.8) as directed in the text. Achieve the same result by replacing $\Gamma(\sigma)$ in (7.8) by its value from equation (7.2). Explain fully how equation (7.8), involving real variables only, can be used to obtain the complex-valued functional equation.

4. Prove equation (4.1).

5. Prove that for all θ

$$35 + 56 \cos \theta + 28 \cos 2\theta + 8 \cos 3\theta + \cos 4\theta \geq 0.$$

Could this be used to prove Theorem 4?

6. Prove
$$\int_0^\infty \frac{x^{\sigma - 1}}{e^x + 1} \, dx = \Gamma(\sigma)\eta(\sigma), \qquad \sigma > 0.$$

7. Prove
$$\int^\infty \frac{x^\sigma}{e^x + 1} \, dx = (1 - 2^{-\sigma}) \int_0^\infty \frac{x^\sigma}{e^x - 1} \, dx, \qquad \sigma > 0.$$

8. What is the order of the trivial zeros of the zeta-function?

9. Show that the function $\Gamma(s/2)\zeta(s)\pi^{-s/2}$ satisfies the functional equation

$$f(s) = f(1 - s).$$

10. Show that

$$\zeta''(\sigma + iy) = O(\log^3 |y|), \qquad |y| \to \infty$$

uniformly in $1 \leq \sigma \leq 2$.

11. By Comrie's computations $N_0(1468) = 1041$. Compare this with Riemann's asymptotic estimate for $N(T)$.   Ans.   1704.

12. Prove that the integral (2.4) represents an analytic function of s for $\sigma > 0$ by use of Theorem 4.2 of Chapter 5. Treat the last integral in (7.6) similarly.

4 · *The Prime Number Theorem*

1. Introduction

As an application of the theory of Dirichlet series we propose to prove in this chapter the "prime number theorem." It says essentially that the nth prime is asymptotic to $n \log n$ as n becomes infinite. This result, long conjectured, was finally proved at the close of the 19th century by J. Hadamard and C. J. de la Vallée Poussin. With some preliminary knowledge of number-theoretic functions in common use the theorem can now be proved in a page or two. But we shall not assume such knowledge. We take rather a more historic approach, proving first a theorem of Tchebychev which was the forerunner of the final theorem. It is hoped that this presentation will give a better appreciation of the difficulties involved. Our proof is a modification of the original one, due largely to E. Landau.

2. The Function $\pi(x)$

As in §3 of Chapter 3 we denote the nth prime by p_n. The symbol $\pi(x)$ is traditionally used for the number of primes $\leq x$.

Definition 2.1.

$$\pi(x) = \sum_{p_k \leq x} 1.$$

By this symbolism we mean that 1 is to be added for each prime $\leq x$. For example, $\pi(7) = 4$, $\pi(5\pi) = 6$, $\pi(p_n) = n$. Thus $\pi(x)$ and p_n are inverse functions. By custom, the one is defined only on the integers, the other is a function of the real variable x.

Theorem 2.1.

$$\pi(\infty) = \infty.$$

This is the familiar fact, known to Euclid, that there are infinitely many primes. Euclid's proof is by contradiction. Suppose that the only primes were $p_1, p_2, \ldots, p_n$. Then every integer greater than p_n must be composite, that is, it must be divisible by p_k for some $k \leq n$. But the integer

$$(2.1) \qquad\qquad p_1 \, p_2 \cdots p_n + 1$$

is not so divisible, and the proof is complete. Note that this proof does not imply that numbers of the form (2.1) are necessarily prime. Thus, if $n = 6$, the number (2.1) is $30031 = (509)(59)$.

We give a second proof of Theorem 2.1, due to G. Pólya, which involves the Fermat numbers

$$(2.2) \qquad\qquad F_n = 1 + 2^{2^n}.$$

None of them is divisible by 2. No two of them have a common factor. For, suppose F_n and F_{n+k} are both divisible by the prime $p > 2$. Then we can show, by use of the elementary algebraic fact that $x^{2m} - 1$ is divisible by $x + 1$, that $F_{n+k} - 2$ is also divisible by p:

$$(2.3) \qquad F_{n+k} - 2 = (2^{2^n})^{2^k} - 1 = (2^{2^n} + 1)N = F_n \, N.$$

Here N is an integer resulting from the algebraic factorization just mentioned ($x = 2^{2^n}$). Since p divides F_n and F_{n+k} it must divide 2 by equation (2.3), a contradiction. Hence there are at least n primes $\leq F_n$, since each Fermat number is either a prime or has prime factors. There can be no duplication in this count, as we proved above. That is,

$$n \leq \pi(2^{2^n} + 1),$$

$$\infty = \pi(\infty).$$

This proof gives the following crude estimates

$$p_n \leqq 2^{2^n} + 1, \qquad n = 1, 2, 3, \ldots,$$

$$A \log \log x \leqq \pi(x), \qquad \text{some } A > 0, \quad 3 \leqq x < \infty.$$

The Fermat numbers are of special interest historically on account of Fermat's conjecture that they are all primes. Euler disproved this by noting that

$$F_5 = 2^{32} + 1 = (641)\,(6700417).$$

We prove next Euler's result that the series of prime reciprocals diverges.

Theorem 2.2.

(2.4) $$\sum_{k=1}^{\infty} \frac{1}{p_k} = \infty.$$

From equation (3.2) of Chapter 3, with $s = 1$, we have

$$\prod_{k=1}^{N} \frac{1}{1 - p_k^{-1}} \geqq \sum_{k=1}^{N} \frac{1}{k}.$$

Since the harmonic series diverges, we have that

(2.5) $$\prod_{k=1}^{\infty} \left(1 - \frac{1}{p_k}\right) = 0.$$

That is, the product (2.5) diverges to zero. If the series (2.4) converged the product (2.5) would converge, so that the theorem is proved. Equation (2.5) is an equally important conclusion for our purposes.

The previous result about p_n must of course have a corollary concerning the inverse function $\pi(x)$. To obtain it, however, we need a lemma from elementary number theory.

Definition 2. $\pi_\rho(x)$ is the number of positive integers $\leqq x$ not divisible by any of the first ρ primes.

For example, $\pi_2(17) = 6$. For, 1, 5, 7, 11, 13, 17 are the only six positive integers $\leqq 17$ not divisible by 2 or 3. Note that the first two primes themeselves must be excluded in the count for $\pi_2(x)$ so that

$$\pi(17) = 7 < 2 + \pi_2(17),$$

and in general

(2.6) $$\pi(x) < \rho + \pi_\rho(x).$$

Lemma 2.3.

$$\pi_\rho(x) = [x]$$

$$- \left[\frac{x}{p_1}\right] - \left[\frac{x}{p_2}\right] - \cdots - \left[\frac{x}{p_\rho}\right]$$

(2.7)
$$+ \left[\frac{x}{p_1 p_2}\right] + \cdots + \left[\frac{x}{p_{\rho-1} p_\rho}\right]$$

$$\cdots$$

$$+ (-1)^\rho \left[\frac{x}{p_1 p_2 \cdots p_\rho}\right].$$

Here the bracket indicates "largest integer $\leq$" and the denominators in the $(k+1)$th row run through the $\binom{\rho}{k}$ products of the first ρ primes taken k at a time. For example,

$$\pi_2(17) = 17 - \left[\frac{17}{2}\right] - \left[\frac{17}{3}\right] + \left[\frac{17}{6}\right] = 17 - 8 - 5 + 2 = 6.$$

We prove this well-known formula by induction on x. It is valid for $0 < x < 1$ when each side of equation (2.7) is zero. Assume it true in the interval $0 < x < N$, where N is an integer not divisible by any of the first ρ primes. It remains true in $0 < x < N+1$ for, as x moves into the larger interval only the first line of formula (2.7) changes; it increases by unity on both sides. Next make the same assumption when N is divisible by just m of the first ρ primes. Then the $(k+1)$th row $(k \leq m)$ adds $(-1)^k \binom{m}{k}$ as x moves into the larger interval. The remaining rows $(k > m)$ do not change. That is, a total of

$$\sum_{k=0}^{m} (-1)^k \binom{m}{k} = (1-1)^m = 0$$

is added, so that (2.7) again remains true in the larger interval. The proof is complete.

Theorem 2.3.

$$\pi(x) = o(x), \qquad x \to \infty.$$

Inequality (2.6) gives a relation between $\pi(x)$ and $\pi_\rho(x)$. Now observe that if the brackets are omitted in formula (2.7) the right-hand side becomes

$$x \prod_{k=1}^{\rho} \left(1 - \frac{1}{p_k}\right).$$

Since dropping a bracket alters an individual term by at most unity and since there are $\binom{\rho}{k}$ terms in the $(k+1)$th row the result is an increase of at most

$$\sum_{k=1}^{\rho} \binom{\rho}{k} = (1+1)^\rho = 2^\rho.$$

It is true that in alternate rows the dropping of brackets *decreases* the sum of the row, but our statement remains true *a fortiori*. Hence

$$\pi(x) < \rho + 2^\rho + x \prod_{k=1}^{\rho} \left(1 - \frac{1}{p_k}\right).$$

Allowing x to become infinite we obtain

$$(2.8) \qquad \overline{\lim_{x \to \infty}} \, \frac{\pi(x)}{x} \leqq \prod_{k=1}^{\rho} \left(1 - \frac{1}{p_k}\right).$$

The left-hand side of (2.8) must be zero since it is ≥ 0 and is independent of ρ, whereas the right-hand side tends to zero as $\rho \to \infty$. Hence $\lim \pi(x)/x = 0$, as we wished to prove. As observed above Theorem 2.3 must be considered a corollary of Theorem 2.2, since it essentially expresses the same result, in terms of the function $\pi(x)$ inverse to p_n. It is rare that a theorem and its corollary describe opposite trends. But this seems to be the situation here, in the following sense. Theorem 2.2 describes the distribution of primes as "rather thick"—not so "sparse" for example as the square numbers ($\sum k^{-2} < \infty$). On the contrary Theorem 2.3 describes the distribution as "rather thin". Thus, the probability of picking a prime from the first n positive integers is $\pi(n)/n$, so that by Theorem 2.3 the probability of picking a prime from *all* the integers is zero (in a precise sense). By use of these two conflicting trends one might conjecture the prime number theorem. It can be shown that the functions

$$f(n) = n \log n, \qquad g(x) = \frac{x}{\log x}$$

are essentially inverse functions (see Exercise 1 of this chapter).

But if p_n behaved like $f(n)$ for large n and $\pi(x)$ like $g(x)$ for large x both the above theorems would be satisfied. We shall later prove the prime number theorem, which states that

$$p_n \sim n \log n \quad \text{or} \quad \pi(x) \sim \frac{x}{\log x}$$

as the variable becomes infinite.

3. The Function $\vartheta(x)$

To make further progress in the study of primes we define a new function, more natural to the study than $\pi(x)$ since it involves *products* of primes rather than sums.

Definition 3.

$$(3.1) \qquad \vartheta(x) = \sum_{p_k \leq x} \log p_k = \int_1^x \log x \, d\pi(x).$$

The notation means that if $p_1, p_2, \ldots, p_m$ are all the primes $\leq x$ then

$$(3.2) \qquad \vartheta(x) = \log \left(\prod_{k=1}^m p_k \right).$$

The Stieltjes integral representation (3.1) follows from the fact that $\pi(x)$ is a step-function with unit jumps at the primes. For example, $\vartheta(10) = 5.347$. A simple numerical observation about the binomial coefficient $\binom{2m}{m}$ now gives us an upper estimate for $\vartheta(x)$.

Theorem 3.1.

$$\vartheta(x) = O(x), \qquad x \to \infty.$$

For,

$$(3.3) \qquad \binom{2m}{m} = \frac{(2m)!}{m!m!} = \frac{(2m)(2m-1) \cdots (m+1)}{1 \cdot 2 \cdots \cdots m}.$$

The numerator on the right of (3.3) is clearly divisible by all the primes between $m+1$ and $2m$, so that the same must be true of the integer on the left:

$$(3.4) \qquad \prod_{m < p_k \leq 2m} p_k \ \bigg| \ \binom{2m}{m}.$$

Here the vertical line is to be read "divides" or "is a factor of". But $\binom{2m}{m} < 2^{2m}$ since it is only one of the (positive) terms in the binomial expansion $(1 + 1)^{2m}$. If m is a power of 2, $m = 2^n$, (3.4) gives

$$(3.5) \qquad \prod_{2^{n-1} < p_k \leq 2^n} p_k < 2^{2^n}.$$

Multiplying together n inequalities (3.5), corresponding to the first n positive integers, gives

$$(3.6) \qquad \prod_{1 < p_k \leq 2^n} p_k < 2^{1+2+4+\cdots+2^n} < 2^{2^{n+1}}.$$

From (3.2) and (3.6)

$$\vartheta(2^n) < 2^n \log 4.$$

Since $\vartheta(x) \in \uparrow$, then for $2^{n-1} < x < 2^n$

$$\frac{\vartheta(x)}{x} < \frac{2^n \log 4}{2^{n-1}} < \log 16.$$

Thus $\vartheta(x)/x$ is bounded above, and our theorem is proved.

This result now gives an improved estimate for $\pi(x)$.

Theorem 3.2.

$$\pi(x) = O\left(\frac{x}{\log x}\right), \qquad x \to \infty.$$

For, from (3.1)

$$\vartheta(x) = \pi(x) \log x - \int_1^x \frac{\pi(t)}{t} \, dt,$$

$$(3.7) \qquad \pi(x) \frac{\log x}{x} = \frac{\vartheta(x)}{x} + \frac{1}{x} \int_1^x \frac{\pi(t)}{t} \, dt.$$

By theorem 2.3 the integrand of (3.7), and hence the last term of (3.7), is bounded above. The same is true of $\vartheta(x)/x$ by Theorem 3.1, so that the result is established.

A second corollary of Theorem 3.1 is that $p_n/(n \log n)$ is bounded away from zero.

Theorem 3.3.

$$p_n > Bn \log n, \qquad \text{some } B > 0, \quad n = 1, 2, 3, \ldots.$$

For, by Theorem 3.2 there exists $B > 0$ such that

$$\pi(x) < \frac{1}{B} \frac{x}{\log x}, \qquad x > 1.$$

Setting $x = p_n$, this becomes

$$n < \frac{1}{B} \frac{p_n}{\log p_n} < \frac{p_n}{B \log n}.$$

4. The Function $\psi(x)$

In order to obtain a lower bound for the rate of growth of $\pi(x)$ we introduce a new function $\psi(x)$, somewhat larger than $\vartheta(x)$. The fact that it is larger enables us to use the same device, involving a binomial coefficient, as was used for $\vartheta(x)$. As we shall see, $\psi(x)$ is not so much bigger than $\vartheta(x)$ as to spoil its usefulness in estimating the rate of growth of $\pi(x)$.

Definition 4.1.

$$\psi(x) = \sum_{p_k^n \leq x} \log p_k.$$

This notation means that we are to add $\log p_k$ for each integer $\leq x$ which is a power of p_k. For example,

$$\psi(10) = 3 \log 2 + 2 \log 3 + \log 5 + \log 7 = \log (2^3 \cdot 3^2 \cdot 5 \cdot 7).$$

Note that in this example $\psi(10)$ is the log of the LCM of the first 10 positive integers. Set

$$U(n) = \text{LCM of } 1, 2, 3, \ldots, n = p_1^{k_1} p_2^{k_2} \cdots p_m^{k_m},$$

where k_j is the largest integer for which $p_j^{k_j} \leq n$ and where no integer $\leq n$ has a larger prime factor than p_m. We thus see that

$$\psi(n) = \log U(n),$$

as in the special case $n = 10$.

Let us prove next two results from elementary number theory.

Lemma 4.1.

$$1. \quad 0 < x, \qquad m = 1, 2, 3, \ldots$$

$$\Rightarrow \qquad \left[\frac{[x]}{m}\right] = \left[\frac{x}{m}\right].$$

Set $[x/m] = n$. By definition of the bracket symbol we have

$$n \leqq \frac{x}{m} < n + 1, \qquad mn \leqq x < mn + m,$$

$$n \leqq \frac{[x]}{m} < n + 1, \qquad \left[\frac{[x]}{m}\right] = n.$$

Lemma 4.2.

$$1. \quad 0 < x$$

$$\Rightarrow \qquad [2x] - 2[x] \leqq 1.$$

For,

$$x < [x] + 1,$$

$$[2x] \leqq 2x < 2[x] + 2, \qquad [2x] - 2[x] < 2.$$

But an integer less than 2 cannot be greater than 1.

Lemma 4.3.

$$1. \quad m = 1, 2, 3, \ldots, \qquad p \text{ is a prime}$$

$$\Rightarrow \qquad \text{The highest power of } p \text{ which divides } m! \text{ is}$$

$$(4.1) \qquad H(p, m!) = \sum_{k=1}^{\infty} \left[\frac{m}{p^k}\right].$$

There are $m_1 = [m/p]$ multiples of p which are $\leqq m$ and their product is $p^{m_1} m_1!$. Repeat this statement when m is replaced by m_1, thus defining $m_2 = [m_1/p]$. By Lemma 4.1 $m_2 = [m/p^2]$. Repeat the statement for m_2, defining m_3, and so on. The required power will be $m_1 + m_2 + m_3 + \ldots$, as stated. All terms of the series after a certain point will be zero since $p^k > m$ for all large k. Hence we may write (4.1) as a finite series as follows:

$$H(p, m!) = \sum_{k=1}^{M} \left[\frac{m}{p_k}\right],$$

where the integer M is determined by the inequalities

(4.2) $$p^M \leqq M < p^{M+1}.$$

We now obtain an estimate from below of the rate of growth of $\psi(x)$.

Theorem 4.3.

$$\psi(x) > Bx, \qquad \text{some } B > 0, \quad 2 \leqq x < \infty.$$

We show first that the binomial coefficient $\binom{2m}{m}$ divides the least common multiple of all positive integers $\leqq 2m$:

(4.3) $$\left. \frac{(2m)!}{m!m!} \right| U(2m) \,.$$

Let p be any prime divisor of $\binom{2m}{m}$. By Lemma 4.3

(4.4) $$H\left(p, \binom{2m}{m}\right) = \sum_{k=1}^{\infty} \left[\frac{2m}{p^k}\right] - 2\left[\frac{m}{p^k}\right]$$

(4.5) $$\leqq \sum_{k=1}^{N} 1 = N,$$

where

(4.6) $$p^N \leqq 2m < p^{N+1}.$$

Here we have used Lemma 4.2 and (4.2). It is to be noted that the second bracket on the right of (4.2) will be zero when k is near N, but this fact does not alter the validity of (4.5). By (4.6) p^N is one of the integers $\leqq 2m$ so that p^N divides $U(2m)$. Thus (4.3) is established, and hence

$$U(2m) \geqq \frac{(m+1)(m+2)\cdots(2m)}{1 \quad 2 \quad \cdots \quad m} \geqq 2^m,$$

$$\psi(2m) \geqq m \log 2.$$

Now if $2m-2 < x < 2m$

$$\frac{\psi(x)}{x} > \frac{\psi(2m-2)}{2m-2} \frac{2m-2}{2m} > \frac{1}{2} \log 2, \qquad m > 2.$$

We thus have a positive lower bound for $\psi(x)/x$ valid for all $x > 2$, as desired.

Since the inequality $p_k{}^n \leqq x$ used in defining $\psi(x)$ is equivalent to $p_k \leqq x^{1/n}$, it is easy to relate $\psi(x)$ to $\vartheta(x)$:

$$\psi(x) = \vartheta(x) + \vartheta(x^{1/2}) + \vartheta(x^{1/3}) + \cdots.$$

This series has all terms zero when $x^{1/k}$ becomes less than 2, so that

$$(4.7) \qquad \psi(x) = \vartheta(x) + \sum_{k=2}^{N} \vartheta(x^{1/k}), \qquad 2^N \leqq x < 2^{N+1}.$$

This relation shows that $\psi(x)$ and $\vartheta(x)$ have the same order of magnitude in the following sense.

Theorem 4.4.

$$\psi(x) = \vartheta(x) + o(x), \qquad x \to \infty.$$

For, from Theorem 3.1 we have

$$(4.8) \qquad \sum_{k=2}^{N} \vartheta(x^{1/k}) \leqq A \sum_{k=2}^{N} x^{1/k} \leqq AN\sqrt{x}, \qquad \text{some } A > 0.$$

But by (4.7) $N \leqq (\log x)/\log 2$, from which we see that the right-hand side of (4.8) $= o(x)$ as $x \to \infty$. This proves the result.

Theorem 4.5.

$$\vartheta(x) > Ax, \qquad \text{some } A > 0, \quad 2 < x < \infty.$$

For, by Theorem 4.3

$$\frac{\vartheta(x)}{x} = \frac{\psi(x)}{x} + o(1) > B + o(1), \qquad \text{some } B > 0.$$

The term $o(1) > -B/2$, for example, for large x, so that the proof is easily completed.

Theorem 4.6.

$$\psi(x) = O(x), \qquad x \to \infty.$$

The proof is similar to the previous one, using Theorems 3.1 and 4.4.

We can now apply our estimates on the growth of $\psi(x)$ and $\vartheta(x)$ to obtain corresponding ones for $\pi(x)$ and p_n.

Theorem 4.7.

$$\pi(x) \log x = \vartheta(x) + o(x), \qquad x \to \infty.$$

This follows from equation (3.1) and from Theorem 2.3. By the latter we conclude that for each $\varepsilon > 0$ there is a number c such that $\pi(t) < \varepsilon t$ when $t > c$. Hence

$$(4.9) \qquad \int_1^x \frac{\pi(t)}{t} \, dt < \int_1^c \frac{\pi(t)}{t} \, dt + \int_c^x \varepsilon \, dt, \qquad \varlimsup_{x \to \infty} \frac{1}{x} \int_1^x \frac{\pi(t)}{t} \, dt \leqq \varepsilon.$$

This implies that the integral on the left of (4.9) is $o(x)$ as $x \to \infty$, and the proof is complete.

Theorem 4.8.

$$(4.10) \qquad \pi(x) > A \frac{x}{\log x}, \qquad \text{some } A > 0, \quad 2 < x < \infty.$$

For, from Theorems 4.5 and 4.7

$$\pi(x) \log x > Bx + o(x) > Bx - \frac{Bx}{2} = \frac{Bx}{2}, \qquad \text{some } B > 0$$

for all large x. The conclusion is trivial for small $x > 2$.

Theorem 4.9.

$$p_n = O(n \log n), \qquad n \to \infty.$$

Replacing x by p_n in (4.10) gives

$$(4.11) \qquad n > A p_n / \log p_n,$$

$$\log n > \log A + \log p_n - \log \log p_n,$$

$$1 > \frac{\log n}{\log p_n} > 1 + \frac{\log A}{\log p_n} - \frac{\log \log p_n}{\log p_n}.$$

Allowing n to become infinite gives

$$(4.12) \qquad \lim_{n \to \infty} \frac{\log n}{\log p_n} = 1,$$

$$(4.13) \qquad \log p_n < 2 \log n, \qquad n > m, \text{ some } m.$$

From (4.11) and (4.13)

$$p_n < \frac{2}{A} n \log n, \qquad n > m.$$

The desired conclusion is now immediate.

The conclusions of §3 and §4 may now be summarized in the following theorem of Tchebychev.

Theorem 4.10. There exist positive numbers A and B such that

$$Bx < \vartheta(x) < Ax,$$

$$Bx < \psi(x) < Ax,$$

$$\frac{Bx}{\log x} < \pi(x) < \frac{Ax}{\log x},$$

$$Bn \log n < p_n < An \log n.$$

While this result gives good estimates for the rate of growth of the four functions involved, it falls short of the prime number theorem, which asserts, for example, that $p_n \sim n \log n$ as $n \to \infty$. The following theorem shows that there are equivalent statements involving the other three functions.

Theorem 4.11. The existence of any one of the following limits implies that of the other three:

$$(4.14) \qquad \lim \frac{p_n}{n \log n} = 1, \quad \lim \frac{\pi(x) \log x}{x} = 1, \quad \lim \frac{\vartheta(x)}{x} = 1,$$

$$\lim \frac{\psi(x)}{x} = 1.$$

In view of Theorems 4.4 and 4.7 it will be sufficient to prove the equivalence of the first two limits. Assume the second and let $x \to \infty$ through the primes. Then

$$(4.15) \qquad \lim_{n \to \infty} \frac{n \log p_n}{p_n} = 1.$$

But this implies the first limit (4.14) by (4.12). Conversely, assume the first limit (4.14) and let x lie between two primes p_n and p_{n+1}, so that

$$(4.16) \qquad \frac{n \log p_n}{p_{n+1}} \leqq \frac{\pi(x) \log x}{x} \leqq (n+1) \frac{\log p_{n+1}}{p_n}.$$

But from (4.12) and the assumed limit it is clear that the two extreme terms of (4.16) tend to unity as $n \to \infty$. This gives the desired conclusion.

5. Five Lemmas

In the proof of the prime number theorem we shall need several preliminary results from analysis. Some of these are already proved in an earlier chapter, but for clarity we restate them here in appropriate context.

Lemma 5.1.

$$(5.1) \qquad \int_{1/2}^{x} \log \frac{x}{t} \, d[t] \sim x, \qquad x \to \infty.$$

Since $t - 1 < [t] \leqq t$, we obtain upper and lower bounds for the integral (5.1), after integration by parts, as follows:

$$x - \frac{1}{2} - \log 2x < \int_{1/2}^{x} \frac{[t]}{t} \, dt < x.$$

When divided by x the two extremes of this continued inequality $\to 1$ as $x \to \infty$, and the lemma is proved.

The next result is the familiar Riemann–Lebesgue theorem. We assume it known for a finite range and prove it for an infinite one.

Lemma 5.2.

$$1. \quad f(x) \in L \, (-\infty, \, \infty)$$

$$\Rightarrow \qquad \lim_{x \to +\infty} \int_{-\infty}^{\infty} e^{ixt} f(t) dt = 0.$$

For, if $R > 0$

$$\varlimsup_{x \to \infty} \left| \int_{-\infty}^{\infty} e^{ixt} f(t)\, dt \right|$$

$$= \varlimsup_{x \to \infty} \left| \int_{-\infty}^{\infty} e^{ixt} f(t)\, dt - \int_{-R}^{R} e^{ixt} f(t)\, dt \right| \leqq \int_{|t| > R} |f(t)|\, dt.$$

Since the left side is a number independent of R and since the right side is $o(1)$ as $R \to \infty$, the proof is complete.

The next result is a special case of Theorem 9.2, Chapter 2.

Lemma 5.3.

$$1. \quad f(s) = \sum_{k=1}^{\infty} \frac{a_k}{k^s}, \qquad \sigma > \sigma_c\,;$$

$$2. \quad a > 0, \qquad a > \sigma_a$$

$$\Rightarrow \qquad \sum_{k \leqq x} a_k \log \frac{x}{k} = \frac{1}{2\pi i} \int_{a - i\infty}^{a + i\infty} \frac{f(s) x^s}{s^2}\, ds.$$

This is equation (9.10) of Chapter 2 when $p = 2$, $\lambda_k = \log k$, $\omega = \log x$. It becomes (9.11) of Chapter 2 for $f(s) = \zeta(s)$.

We shall need some properties of the function $\zeta(s) + (\zeta'(s)/\zeta(s))$, essentially proved in Chapter 3.

Lemma 5.4.

$$1. \quad f(s) = -\zeta(s) - \frac{\zeta'(s)}{\zeta(s)}$$

$$\Rightarrow \qquad \text{A.} \quad f(s) \text{ is analytic for } \sigma \geqq 1,$$

$$\text{B.} \quad f(\sigma + iy) = O(\log^9 |y|), \qquad |y| \to \infty$$

uniformly in $1 \leqq \sigma \leqq 2$.

These were essentially contained in §8 of Chapter 3. As we shall see in more detail in Chapter 8 a Tauberian theorem is one which draws a conclusion about the limit of a function from a hypothesis about the limit of its average. It is just such a result which we now state.

Lemma 5.5.

$$1. \quad xf(x) \in \uparrow, \qquad 1 \leq x < \infty;$$

$$2. \quad \int_1^x f(t)\, dt \sim ax, \qquad x \to +\infty$$

$$\Rightarrow \qquad f(x) \sim a, \qquad x \to +\infty.$$

Let δ be an arbitrary positive number. By hypothesis 2,

$$(5.2) \qquad \int_1^x f(t)\, dt = ax + o(x), \qquad x \to \infty,$$

$$\int_x^{x(1+\delta)} f(t)\, dt = ax + a\delta x - ax + o(x) = a\delta x + o(x), \qquad x \to \infty.$$

By hypothesis 1 we have for $1 < x \leq t \leq x(1 + \delta)$,

$$xf(x) \leq tf(t) \leq x(1 + \delta)f(x + \delta x),$$

$$\frac{x}{t}f(x) \leq f(t) \leq \frac{x}{t}(1 + \delta)f(x + \delta x).$$

But

$$\frac{1}{1 + \delta} \leq \frac{x}{t} \leq 1,$$

so that

$$\frac{f(x)}{1 + \delta} \leq f(t) \leq (1 + \delta)f(x + \delta x),$$

$$\frac{\delta x f(x)}{1 + \delta} = \int_x^{x(1+\delta)} f(t)\, dt \leq \delta x(1 + \delta)f(x + \delta x).$$

Now using (5.2) we obtain

$$\frac{\delta f(x)}{1 + \delta} \leq a\delta + o(1) \leq \delta(1 + \delta)f(x + \delta x),$$

$$\overline{\lim_{x \to \infty}} f(x) \leq a(1 + \delta),$$

$$\underline{\lim_{x \to \infty}} f(x + \delta x) = \underline{\lim_{x \to \infty}} f(x) \geq \frac{a}{1 + \delta}.$$

Allowing δ to approach zero, we see that

$$a \leq \underline{\lim} \, f(x) \leq \overline{\lim} \, f(x) \leq a,$$

and the lemma is proved.

6. Background and Proof of the Prime Number Theorem

Before proving the main theorem let us sketch the motivating idea. By use of Euler's product one immediately obtains a double series expansion for $\log \zeta(s)$,

$$\log \zeta(s) = - \sum_{k=1}^{\infty} \log (1 - p_k^{-s}), \qquad \sigma > 1$$

$$= \sum_{k=1}^{\infty} \sum_{n=1}^{\infty} \frac{1}{n} \frac{1}{p_k^{ns}}, \qquad \sigma > 1,$$

which converges, by the very way in which it was obtained, for $\sigma > 1$. Rearranging this double series as a simple series expresses $\log \zeta(s)$, and hence its derivative, as a Dirichlet series,

$$(6.1) \qquad -\frac{\zeta'(s)}{\zeta(s)} = \sum_{k=1}^{\infty} \frac{\Lambda(k)}{k^s},$$

where

$$\Lambda(k) = \log p, \qquad k = p^n$$

$$= 0, \qquad k \neq p^n.$$

That is, $\Lambda(k)$ is zero when k is not the power of a prime, is $\log p$ when k is any power of the prime p. Observing that $\psi(x)$ is a step-function with jump $\Lambda(k)$ at the integer $k = p^m$, we also see that

$$(6.2) \qquad -\frac{\zeta'(s)}{\zeta(s)} = \int_1^{\infty} \frac{d\psi(t)}{t^s}, \qquad \sigma > 1,$$

$$\psi(x) = \sum_{k \leq x} \Lambda(k).$$

From Theorem 9.1 of Chapter 2, x not an integer,

$$(6.3) \qquad \psi(x) = \frac{-1}{2\pi i} \int_{a-i\infty}^{a+i\infty} \frac{\zeta'(s)}{\zeta(s)} \frac{x^s}{s} \, ds, \qquad a > 1.$$

If one had sufficient information about $\zeta(s)$ in the critical strip, one might hope to change the constant a in (6.3) to $b < 1$, thus introducing the residue x of the integrand at $s = 1$:

$$(6.4) \qquad \psi(x) = x - \frac{1}{2\pi i} \int_{b-i\infty}^{b+i\infty} \frac{\zeta'(s)}{\zeta(s)} \frac{x^s}{s} \, ds.$$

If this new integral could be shown to be $o(x)$ as $x \to \infty$, we would have $\psi(x) \sim x$, as desired.

In the original proofs by Hadamard and de la Vallée Poussin, the procedure outlined above was generally followed. Due to the paucity of information about $\zeta(s)$ in the critical strip the line $\sigma = b$ in (6.4) had to be replaced, as path of integration, by a curve having the line $\sigma = 1$ as asymptote. In the modified proof (due to E. Landau, G. H. Hardy, A. E. Ingham, *et al*) which we shall give, there will be essentially three changes in the above procedure:

 A. Replace $-\zeta'/\zeta$ of (1) by $f(s)$ of Lemma 5.4,
 B. Replace s in the denominator of (6.3) by s^2,
 C. Replace b by 1 in (6.4).

The last alteration is possible since $f(s)$ is analytic on $\sigma = 1$. Cauchy's residue theory is replaced, as a tool, by the Riemann–Lebesgue theorem.

By Theorem 4.11 the prime number theorem is equivalent to the following theorem.

Theorem 6.

$$\psi(x) \sim x, \qquad x \to \infty.$$

In view of equation (6.2)

$$-\frac{\zeta'(s)}{\zeta(s)} - \zeta(s) = f(s) = \int_{1/2}^{\infty} \frac{1}{t^s} \, d(\psi(t) - [t]), \qquad \sigma > 1.$$

By Lemma 5.3

$$\sum_{k \leq x} (\Lambda(k) - 1) \log \frac{x}{k} = \frac{1}{2\pi i} \int_{a-i\infty}^{a+i\infty} f(s) \frac{x^s}{s^2} \, ds, \qquad a > 1,$$

$$(6.5) \quad \int_{1/2}^{x} \log \frac{x}{t} \, d\psi(t) = \int_{1/2}^{\lambda} \log \frac{x}{t} \, d[t] + \frac{x^a}{2\pi} \int_{-\infty}^{\infty} \frac{f(a + iy)}{(a + iy)^2} x^{iy} \, dy.$$

By Lemma 5.4, for $1 \leq a \leq 2$ there exists a positive constant M such that

$$(6.6) \qquad \int_{|y| \geq 1} \frac{f(a + iy)}{(a + iy)^2} x^{iy} \, dy \ll M \int_{|y| \geq 1} \frac{\log^9 |y|}{1 + y^2} \, dy < \infty.$$

By Lebesgue's limit theorem or by uniform convergence the last integral (6.5) approaches

$$\int_{-\infty}^{\infty} \frac{f(1 + iy)}{(1 + iy)^2} x^{iy} \, dy$$

as $a \to 1-$, and the latter integral is $o(1)$ as $x \to +\infty$ by Lemma 5.2. Now an appeal to Lemma 5.1 and equation (6.5) shows that

$$(6.7) \qquad \int_{1/2}^{x} \log \frac{x}{t} \, d\psi(t) = \int_{1/2}^{x} \frac{\psi(t)}{t} \, dt = x + o(x), \qquad x \to \infty.$$

Since $\psi(t) \in \uparrow$, the hypotheses of Lemma 5.5 are satisfied for $f(t) = \psi(t)/t$, and we conclude that $\psi(x) \sim x$, as was to be proved.

7. Further Developments

The nonvanishing of $\zeta(1 + iy)$ was an essential feature in the proof of Theorem 6. Just how essential may be judged by the following theorem.

Theorem 7.

    1. $\quad \psi(x) \sim x, \qquad x \to \infty \Rightarrow \zeta(1 + iy) \neq 0, \qquad -\infty < y < \infty.$

To obtain a contradiction assume that $\zeta(s)$ has a root of order m at $s_0 = 1 + iy_0$, $y_0 \neq 0$. Since the logarithmic residue at s_0 is then equal to m we have

$$\lim_{s \to s_0} \frac{\zeta'(s)}{\zeta(s)} (s - s_0) = m.$$

From equation (6.2) we have, after integration by parts, that

$$-\frac{\zeta'(s)}{\zeta(s)} = \frac{s}{s - 1} + s \int_{1}^{\infty} \frac{\psi(t) - t}{t^{s+1}} \, dt, \qquad \sigma > 1.$$

In particular, take $s = \sigma + iy_0$, $s - s_0 = \sigma - 1$. For an arbitrary $\varepsilon > 0$ there exists by hypothesis 1 an R such that $|\psi(t) - t| < \varepsilon t$ when $t > R$. Hence

$$\left|\frac{\zeta'(s)}{\zeta(s)}\right|(\sigma - 1) \leqq \left|\frac{s}{s - 1}\right|(\sigma - 1) + |s|(\sigma - 1)\int_1^R \frac{\psi(t) + t}{t^2}\, dt$$

$$+ |s|(\sigma - 1)\varepsilon \int_R^\infty \frac{dt}{t^\sigma},\ \lim_{s \to s_0}\left|\frac{\zeta'(s)}{\zeta(s)}\right|(\sigma - 1) \leqq |s_0|\varepsilon = \varepsilon(1 + y_0^2)^{1/2}.$$

This gives a contradiction if ε is chosen $< m/(1 + y_0^2)^{1/2}$; it shows in fact that the limit in question is zero rather than m. This theorem is of interest in view of a recent "elementary" proof of the prime number theorem (by A. Selberg and P. Erdös) which does not involve complex variables. For an exposition of the elementary proof see N. Levinson [1969, p. 225].

The logarithmic integral

$$\text{li}\,(x) = \lim_{\delta \to 0}\left(\int_0^{1-\delta} \frac{dt}{\log t} + \int_{1+\delta}^x \frac{dt}{\log t}\right) = 1.04 \cdots + \int_2^x \frac{dt}{\log t},$$

being asymptotically equivalent to $x/\log x$,

$$\lim_{x \to \infty} \frac{\log x}{x}\, \text{li}\, x = 1,$$

may also be used as an approximation to $\pi(x)$. In fact, C. F. Gauss proposed this approximation as a result of his examination of tables of primes. It is interesting to compare these two approximations with $\pi(x)$ itself for two large values of x for which $\pi(x)$ is known exactly.

x	$\pi(x)$	$\dfrac{x}{\log x}$	$\text{li}\,(x)$	$\dfrac{\pi(x)}{\text{li}\,(x)}$
10^7	664,579	620,421	664,918	0.999490
10^9	50,847,478	48,254,940	50,849,235	0.999965

Observe that $\pi(x) < \text{li}\,(x)$ for these two values of x. Indeed this inequality holds for all x for which $\pi(x)$ has been computed. Yet Littlewood proved in 1914 that the reverse inequality holds for infinitely many values of x. S. Skewes proved in 1937 that this reverse inequality holds for some value of x less than x_0 (see Skewes [1955]),

$$x_0 = 10^{10^{10^{10^3}}}.$$

The order of magnitude of the difference $\pi(x) - \text{li}\,(x)$ has been studied. Littlewood proved that

$$\pi(x) = \text{li}\,(x) + O(x \exp[-a(\log x \log \log x)^{1/2}]), \qquad x \to \infty$$

for some positive number a. If the Riemann hypothesis is assumed a better result can be proved:

$$\pi(x) = \text{li}\,(x) + O(\sqrt{x} \log x), \qquad x \to \infty.$$

We conclude this section with the statement of an interesting result, conjectured by L. Euler and proved by von Mangoldt. It involves the Möbius function $\mu(n)$:

$$\mu(1) = 1,$$

$$\mu(n) = (-1)^q, \qquad n = p_1 p_2 \ldots p_q, \quad (p_j \text{ are distinct primes}),$$

$$\mu(n) = 0, \qquad \text{otherwise.}$$

It is an elementary result that for $\sigma > 1$

$$(7.1) \qquad \frac{1}{\zeta(s)} = \sum_{n=1}^{\infty} \frac{\mu(n)}{n^s} = 1 - \frac{1}{2^s} - \frac{1}{3^s} - \frac{1}{5^s} + \frac{1}{6^s} - \frac{1}{7^s} + \frac{1}{10^s} - \cdots.$$

It is a much deeper result that this series still converges for $s = 1$ and has the value zero there. It is of interest to us here because E. Landau has shown that the convergence of (7.1) at $s = 1$ is "equivalent" to the prime number theorem.

8. Summary

By elementary means we have found the correct order of magnitude of $\pi(x)$, the number of primes $\leq x$:

$$(8.1) \qquad \frac{Bx}{\log x} < \pi(x) < \frac{Ax}{\log x}, \qquad 2 \leq x < \infty,$$

where A and B are positive constants. The essential device for obtaining the upper bound (8.1) resulted from the observation that the product of all the primes between an integer m and its double divides the binomial coefficient $\binom{2m}{m}$. This same binomial coefficient, in turn, is a divisor of the least common multiple of all positive integers $\leqq x$, and this remark provided the lower bound (8.1).

By relating $\pi(x)$ to the zeta-function of Riemann we obtained the exact asymptotic behavior of $\pi(x)$, in the sense that

$$\pi(x) \sim \frac{x}{\log x}, \qquad x \to \infty.$$

The essential property of $\zeta(s)$ used was that

$$\zeta(1 + iy) \neq 0, \qquad -\infty < y < \infty.$$

A clear and brief summary about the distribution of primes is to be found in G. H. Hardy [1929], his 1928 Gibbs lecture.

EXERCISES

1. Assuming $\pi(x) \sim x/\log x$, prove $\log p_n \sim p_n/n$. By taking logarithms of the latter relation (permissible?) show that $\log p_n \sim \log n$. Thus prove, without use of Theorem 4.10 that

$$\pi(x) \sim \frac{x}{\log x} \Rightarrow p_n \sim n \log n.$$

2. Prove the converse conclusion of Lemma 5.5 without hypothesis 1:

$$f(\infty) = a \Rightarrow \int_1^x f(t) \, dt \sim ax, \qquad x \to \infty.$$

    Of course, $f(x) \in L$ in $(1, R)$ for every $R > 1$.

3. By consideration of $f(x) = 1 + \cos x$ show that the first hypothesis of Lemma 5.5 cannot be omitted.

4. Bertand's conjecture asserts that there is a prime between n and $2n$, $n = 2, 3, 4, \ldots$. Prove it if

$$\frac{0.8\,x}{\log x} < \pi(x) < \frac{1.2\,x}{\log x}, \qquad x \geqq 64.$$

5. $\quad \lim_{n \to \infty} [U(n)]^{1/n} = ?, \qquad U(n) = \text{LCM} \, (1, 2, \ldots, n).$

6. Find σ_a for $\displaystyle\sum_{k=1}^{\infty} U(k) \, e^{-ks}$.

7. Prove

$$\sum_{k=1}^{n} \frac{1}{p_k} \sim \log \log n, \qquad n \to \infty.$$

[Hint:

$$(1 - \varepsilon)k \log k < p_k < (1 + \varepsilon)k \log k, \qquad \text{large } k$$

and

$$\sum \frac{1}{k \log k}$$

may be estimated by the analogous integral.]

8. Prove

$$\sum_{k=1}^{n} \frac{\log p_k}{p_k} \sim \log n, \qquad n \to \infty.$$

9. Prove Theorem 4.6.

10. Prove Theorem 6 by use of the integral

$$\int_{a-i\infty}^{a+i\infty} \frac{f(s)x^s}{s^3} \, ds.$$

5 *The Laplace Transform*

1. Introduction

We turn now to a particular transform of the type described as case A in the first sentence of Chapter 2. The kernel $G(s, t)$ is here e^{-st}, so that the transform is

$$(1.1) \qquad f(s) = \int_0^\infty e^{-st}\phi(t)\, dt$$

and is the continuous analog of a Dirichlet series. We shall derive a set of theorems and formulas for (1.1) quite analogous to the corresponding ones in Chapter 2. However, to see the correspondence more clearly it is convenient to replace (1.1) by another,

$$(1.2) \qquad f(s) = \int_0^\infty e^{-st}\, d\alpha(t),$$

where the integral is now a Stieltjes integral. For the student unfamiliar with such integrals it is suggested that the basic facts may be quickly learned by reading the first few sections of Chapter V, Widder [1961]. Indispensable are the existence theorem, integration by parts, and the

special cases when the integrator function is a step-function or is absolutely continuous. If $\alpha(t)$ in (1.2) is a step-function, that transform reduces to a Dirichlet series; if

$$\alpha(t) = \int_0^t \phi(y)\, dy,$$

(1.2) reduces to (1.1). The greater flexibility of the "Laplace–Stieltjes"
integral (1.2) is thus apparent. As we shall see, it also reduces under
suitable conditions to type (1.1) after integration by parts:

$$f(s) = s \int_0^\infty e^{-st}\alpha(t)\, dt.$$

2. Definitions and Examples

We begin with the following definition.

Definition 2.1. The integral

$$(2.1) \qquad f(s) = \int_0^\infty e^{-st}\, d\alpha(t) = \lim_{R \to \infty} \int_0^R e^{-st}\, d\alpha(t),$$

where $\alpha(t)$ is of bounded variation in $0 \leq t \leq R$ for every positive R, is
called a Laplace–Stieltjes integral.

We automatically assume that $\alpha(t)$ is a complex-valued function
which satisfies the condition of the definition and that $\alpha(0) = 0$ when
the integral (2.1) is written. In particular, if $\alpha(t)$ is absolutely continuous,

$$(2.2) \qquad \alpha(t) = \int_0^t \phi(u)\, du,$$

then (2.1) takes the classical form

$$(2.3) \qquad f(s) = \int_0^\infty e^{-st}\phi(t)\, dt.$$

We shall refer to either equation (2.1) or (2.3), as a Laplace transform. If $\alpha(t)$ is a suitable step-function, (2.1) reduces to a Dirichlet
series.

In §2 of Chapter 1 we gave a list of examples of Laplace transforms. A useful Laplace–Stieltjes integral is

$$e^{-\lambda s} = \int_0^\infty e^{-st}\, dU(t-\lambda), \qquad 0 \leq \lambda < \infty,$$

where

$$U(0) = 0, \qquad U(t) = 1, \qquad t > 0.$$

Definition 2.2. The integral

$$(2.4) \qquad f(s) = \int_{-\infty}^\infty e^{-st}\, d\alpha(t) = \int_0^\infty e^{-st}\, d\alpha(t) + \int_0^\infty e^{st}\, d[-\alpha(-t)]$$

is called a bilateral Laplace–Stieltjes integral.

Again, if $\alpha(t)$ is defined by (2.2), then (2.4) takes the form

$$(2.5) \qquad f(s) = \int_{-\infty}^\infty e^{-st}\phi(t)\, dt.$$

If we set $t = e^{-x}$, (2.5) becomes a Mellin transform:

$$f(s) = \int_0^\infty x^{s-1}\phi(e^{-x})\, dx.$$

Examples are:

A. $\displaystyle \Gamma(s) = \int_{-\infty}^\infty e^{-st}\exp(-e^{-t})\, dt = \int_0^\infty x^{s-1}e^{-x}\, dx, \qquad \sigma > 0.$

B. $\displaystyle \frac{\pi}{2\sin\dfrac{\pi s}{2}} = \int_0^\infty \frac{x^{s-1}}{x^2+1}\, dx, \qquad 0 < \sigma < 2.$

C. $\displaystyle \frac{\pi}{2\cos\dfrac{\pi s}{2}} = \int_0^\infty x^{s-1}\,\frac{x}{x^2+1}\, dx, \qquad -1 < \sigma < 1.$

D. $\displaystyle \Gamma(s)\zeta(s) = \int_0^\infty \frac{x^{s-1}}{e^x-1}\, dx, \qquad \sigma > 1.$

E. $\displaystyle \Gamma(s)\sin\frac{\pi s}{2} = \int_0^\infty x^{s-1}\sin x\, dx, \qquad -1 < \sigma < 1.$

F. $\Gamma(s) \cos \dfrac{\pi s}{2} = \displaystyle\int_0^\infty x^{s-1} \cos x \, dx, \qquad 0 < \sigma < 1.$

G. $\exp s^2 c = \displaystyle\int_{-\infty}^\infty e^{-st} k(t, c) \, dt, \qquad c > 0, \quad -\infty < \sigma < \infty,$

where

$$k(x, c) = (4\pi c)^{-1/2} \exp\left[\frac{-x^2}{(4c)}\right].$$

See Erdélyi [1954, p. 344], for example.

3. Convergence

As might be expected from experience with Dirichlet series the region of convergence of a Laplace integral is a half-plane. Let us prove first the following result.

Lemma 3.

(3.1)   1. $\left| \displaystyle\int_b^t \exp\left(-s_0 u\right) d\alpha(u) \right| \le M, \qquad 0 \le b \le t < \infty$

$\Rightarrow \quad \left| \displaystyle\int_b^\infty e^{-st} \, d\alpha(t) \right| \le M \left| \dfrac{s - s_0}{\sigma - \sigma_0} \right| \exp[-b(\sigma - \sigma_0)], \qquad \sigma > \sigma_0.$

For, if we set $\beta(t)$ equal to the integral (3.1) we have

$$\int_b^R e^{-st} \, d\alpha(t) = \int_b^R \exp[-(s - s_0)t] \, d\beta(t)$$

$$= \beta(R) \exp[-(s - s_0)R]$$

$$+ (s - s_0) \int_b^R \exp[-(s - s_0)t] \, \beta(t) \, dt.$$

Since $\beta(R)$ is bounded by hypothesis

$$\lim_{R \to \infty} \beta(R) \exp\left[-(s - s_0)R\right] = 0, \qquad \sigma > \sigma_0,$$

and

(3.2) $\displaystyle\int_b^\infty e^{-st} \, d\alpha(t) = (s - s_0) \int_b^\infty \exp[-(s - s_0)t]\beta(t) \, dt$

if either integral converges. But

$$\int_b^\infty \exp[-(s-s_0)t]\beta(t)\,dt$$

$$\ll \int_b^\infty \exp[-(\sigma-\sigma_0)t]M\,dt = \frac{M\exp[-b(\sigma-\sigma_0)]}{\sigma-\sigma_0}, \qquad \sigma > \sigma_0.$$

The conclusion of the lemma is now immediate. Note that integration by parts of the integral on the left of (3.2), which may be only conditionally convergent, has replaced it by an absolutely convergent one.

Theorem 3.

$$(3.3) \qquad\qquad 1. \quad \int_0^\infty \exp(-s_0 t)\,d\alpha(t) \quad \text{converges}$$

$$\Rightarrow \qquad\qquad \int_0^\infty e^{-st}\,d\alpha(t) \quad \text{converges for } \sigma > \sigma_0.$$

Since inequality (3.1) with $b=0$ is an immediate consequence of (3.3), the proof is evident. This theorem enables one to define and show the existence of an *abscissa of convergence* σ_c, just as for Dirichlet series. Since Theorem 3 is trivially true if convergence is replaced by absolute convergence throughout, we may also infer the existence of σ_a, the *abscissa of absolute convergence*. The two abscissas need not be equal, as the examples of Chapter 2 show. Another example of classical form is

$$\int_0^\infty e^{-st}e^t \sin e^t\,dt = \int_1^\infty \frac{\sin x}{x^s}\,dx,$$

for which $\sigma_c = 0$, $\sigma_a = 1$.

Corollary 3. If

$$(3.4) \qquad\qquad \int_{-\infty}^\infty e^{-st}\,d\alpha(t)$$

converges for $s = \sigma_0 + i\tau_0$ and for $s = \sigma_1 + i\tau_1$ with $\sigma_0 < \sigma_1$, then it converges for all s in the strip $\sigma_0 < \sigma < \sigma_1$.

For the bilateral Laplace integral one defines two abscissas of convergence $\sigma_c' \leqq \sigma_c''$. In examples A, B, C of §2 they are $(0, \infty)$, $(0, 2)$, and $(-1, 1)$, respectively.

4. Uniform Convergence

We show next that a Laplace integral converges uniformly in certain Stolz regions.

Theorem 4.1. If

$$(4.1) \qquad \int_0^\infty e^{-st}\, d\alpha(t)$$

converges at s_0, it converges uniformly in $\mathrm{St}(s_0)$.

Given $\varepsilon > 0$, we must show that there exists R_0, independent of s in

$$\mathrm{St}(s_0) = \left\{ s \,\middle|\, |\arg(s - s_0)| \leqq a < \frac{\pi}{2}\,;\, \sigma > \sigma_0 \right\}$$

such that when $R > R_0$

$$(4.2) \qquad \left| \int_R^\infty e^{-st}\, d\alpha(t) \right| \leqq \varepsilon.$$

We choose R_0 so that when $t > R > R_0$

$$|\beta(t)| = \left| \int_R^t \exp(-s_0 u)\, d\alpha(u) \right| \leqq \varepsilon \cos a.$$

This is possible by Cauchy's criterion for the existence of a limit. Clearly R_0 has no dependence on s in $\mathrm{St}(s_0)$. Then for $\sigma > \sigma_0$ we have by Lemma 3

$$(4.3) \qquad \left| \int_R^\infty e^{-st}\, d\alpha(t) \right| \leqq \varepsilon\, \frac{|s - s_0|}{\sigma - \sigma_0}\, \cos a.$$

Since

$$|s - s_0|/(\sigma - \sigma_0) = \sec[\arg(s - s_0)] \leqq \sec a$$

in $\mathrm{St}(s_0)$, (4.2) follows from (4.3).

By use of Weierstrass's theorem we have at once the following theorem.

Theorem 4.2. If

$$f(s) = \int_{-\infty}^{\infty} e^{-st}\, d\alpha(t), \qquad \sigma_c' < \sigma < \sigma_c'',$$

then $f(s)$ is analytic there, and for $p = 1, 2, 3, \ldots$

$$f^{(p)}(s) = \int_{-\infty}^{\infty} e^{-st}(-t)^p\, d\alpha(t), \qquad \sigma' < \sigma < \sigma_c''.$$

Of course, this theorem applies to the unilateral Laplace integral as a special case.

5. Formulas for σ_c and σ_a

As in the corresponding derivation of σ_c for Dirichlet series we prove two preliminary results.

Lemma 5.1.

$$1. \quad \alpha(t) = O(e^{\alpha t}), \qquad t \to \infty$$

$$\Rightarrow \qquad \int_0^{\infty} e^{-st}\, d\alpha(t) \text{ converges for } \sigma > \alpha.$$

For,

$$\int_0^R e^{-st}\, d\alpha(t) = \alpha(R)e^{-sR} + s \int_0^R e^{-st}\alpha(t)\, dt.$$

Since for $\sigma > \alpha$

$$\alpha(R)e^{-sR} = O(e^{-\sigma R + \alpha R}) = o(1), \qquad R \to \infty,$$

we have

$$(5.1) \qquad \int_0^{\infty} e^{-st}\, d\alpha(t) = s \int_0^{\infty} e^{-st}\alpha(t)\, dt$$

whenever either integral converges. But

$$\int_0^{\infty} e^{-st}\alpha(t)\, dt \ll \int_0^{\infty} e^{-\sigma t}Me^{\alpha t}\, dt, \qquad \text{some } M.$$

The dominant integral converges for $\sigma > \alpha$, and the proof is complete. Note that the integration by parts used in (5.1) replaces a (conditionally) convergent integral by an absolutely convergent one. It is to be observed that condition 2 of Lemma 6.1, Chapter 2, is not needed here.

Lemma 5.2.

$$1. \quad \int_0^\infty e^{-\alpha t}\, d\alpha(t) \quad \text{converges,} \qquad \text{some } \alpha > 0$$

$$\Rightarrow \qquad \alpha(t) = O(e^{\alpha t}), \qquad t \to \infty.$$

Set

$$\beta(t) = \int_0^t e^{-\alpha u}\, d\alpha(u), \qquad 0 \leq t < \infty.$$

Then

$$\alpha(t) = \int_0^t d\alpha(u) = \int_0^t e^{\alpha u}\, d\beta(u)$$

$$= \beta(t)e^{\alpha t} - \alpha \int_0^t \beta(u)e^{\alpha u}\, du.$$

Since $\beta(t) = O(1)$, $t \to \infty$, there exists $M > 0$ such that

$$|\alpha(t)| \leq M e^{\alpha t} + \alpha M \int_0^t e^{\alpha u}\, du \leq 2M e^{\alpha t}.$$

This completes the proof. Note that this lemma is false for $\alpha < 0$. For, if

$$\alpha(t) = \int_0^t e^{-u}\, du,$$

the integral

$$\int_0^\infty e^{-st}\, d\alpha(t) = \int_0^\infty e^{-st}e^{-t}\, dt$$

converges for $\sigma > -1$. But $\alpha(t) \neq O(e^{\alpha t})$ for any negative α since $\alpha(\infty) = 1$.

We can now obtain easily a formula for σ_c.

Theorem 5.1.

$$
(5.2) \qquad 1. \quad \varlimsup_{t \to \infty} \frac{\log |\alpha(t)|}{t} = \alpha \qquad (\text{or } +\infty);
$$

$$
2. \quad \alpha > 0
$$

$$
(5.3) \quad \Rightarrow \qquad \sigma_c \text{ for } \int_0^\infty e^{-st}\, d\alpha(t) \quad \text{is } \alpha \qquad (\text{or } +\infty).
$$

We prove only the case $\alpha < \infty$. If $\varepsilon > 0$, then hypothesis 1 gives

$$
\alpha(t) = O(e^{(\alpha+\varepsilon)t}), \qquad t \to \infty.
$$

By Lemma 5.1 the integral (5.3) converges for $\sigma > \alpha + \varepsilon$, every $\varepsilon > 0$, and hence $\sigma > \alpha$. Suppose next that (5.3) converged for some point to the left of the line $\sigma = \alpha$. There would then exist a point $s = \beta > 0$, $\beta < \alpha$, where (5.3) would also converge, and by Lemma 5.2

$$
\alpha(t) = O(e^{\beta t}), \qquad t \to \infty.
$$

That is

$$
|\alpha(t)| \leq M e^{\beta t}, \qquad \text{some } M, \quad 0 \leq t < \infty,
$$

$$
\varlimsup_{t \to \infty} \frac{\log|\alpha(t)|}{t} = \alpha \leq \beta < \alpha.
$$

The contradiction shows that (5.3) must diverge for $\sigma < \alpha$, as stated.

The same argument, as that used in §6 of Chapter 2, shows that if σ_c is known in advance to be positive, then it is given by the limit superior (5.2). If $\phi(t) \in L$ in $(0, R)$ for every $R > 0$, then σ_c for the integral

$$
\int_0^\infty e^{-st}\, \phi(t)\, dt
$$

is given by

$$
(5.4) \qquad \sigma_c = \varlimsup_{t \to \infty} \frac{1}{t} \log \left| \int_0^t \phi(u)\, du \right|
$$

if either side is positive. For σ_a replace $\phi(u)$ by $|\phi(u)|$ in (5.4) and $|\alpha(t)|$ by $u(t)$ in (5.2), where $u(t)$ is the total variation of $\alpha(x)$ in $0 \leq x \leq t$.

We give without proof the formulas for $\sigma_c{}'$ and $\sigma_c{}''$ of the bilateral transform (2.4):

$$\sigma_c'' = \overline{\lim_{t \to \infty}} \; \frac{\log|\alpha(t)|}{t}, \qquad \sigma_c' = \underline{\lim_{t \to -\infty}} \; \frac{\log|\alpha(t)|}{t},$$

valid when $0 \neq \sigma_c'' > \sigma_c' \neq 0$.

A consequence of Lemma 5.2 of frequent use follows.

Theorem 5.2.

$$1. \quad f(s) = \int_0^\infty e^{-st} \, d\alpha(t), \qquad \sigma > \sigma_c;$$

(5.5) $\quad \Rightarrow \quad$
$$f(s) = s \int_0^\infty e^{-st} \alpha(t) \, dt, \qquad \sigma > \sigma_c, \quad \sigma > 0,$$

the integral (5.5) converging absolutely.

For, let s_0 be arbitrary with $\sigma_0 > 0$ and $\sigma_0 > \sigma_c$. By Lemma 5.2

(5.6) $$\alpha(t) = O(\exp(\sigma_0 t/2)) \qquad t \to \infty.$$

Hence

$$\alpha(t) \exp(-s_0 t) = O(\exp(-\sigma_0 t/2)) = o(1), \qquad t \to \infty,$$

and integration by parts gives

$$f(s_0) = s_0 \int_0^\infty \exp(-s_0 t)\alpha(t) \, dt.$$

This integral converges absolutely by virtue of (5.6), and the proof is complete.

6. Behavior on Vertical Lines

A function represented by a Laplace integral behaves on vertical lines in much the same way as one which is the sum of a Dirichlet series.

Theorem 6.

(6.1) $$1. \quad f(s) = \int_0^\infty e^{-st} \, d\alpha(t), \qquad \sigma > \sigma_c;$$

$$2. \quad \delta > 0$$

$\Rightarrow$
$$f(\sigma + i\tau) = o(|\tau|), \qquad |\tau| \to \infty$$

uniformly in $\sigma_c + \delta \leqq \sigma < \infty$.

Choose $s_0 = \sigma_c + (\delta/2)$. Since (6.1) converges at s_0 there exists a constant M such that

$$\left| \int_0^x \exp(-s_0 t)\, d\alpha(t) \right| \leq M, \qquad 0 \leq x < \infty,$$

and for any $R > 0$

$$\left| \int_R^x \exp(-s_0 t)\, d\alpha(t) \right| \leq 2M, \qquad R \leq x < \infty.$$

Hence by Lemma 3 we have for $\sigma > \sigma_0$ and any $R > 0$

$$|f(s)| \leq \int_0^R \exp(-\sigma_0 t)|d\alpha(t)| + 2M\, \frac{|s - s_0|}{\sigma - \sigma_0}\, \exp[-R(\sigma - \sigma_0)].$$

For $\sigma \geq \sigma_c + \delta$

$$(6.2) \qquad \frac{|f(\sigma + i\tau)|}{|\tau|} \leq \frac{1}{|\tau|} \int_0^R \exp(-\sigma_0 t)|d\alpha(t)| + 2M\left(\frac{1}{|\tau|} + \frac{2}{\delta}\right)e^{-R\delta/2}.$$

The right-hand side of (6.2) is independent of σ, so that the uniformity will follow automatically if it can be made arbitrarily small by choice of R and $|\tau|$. But

$$\varlimsup_{|\tau| \to \infty} \frac{|f(\sigma + i\tau)|}{|\tau|} \leq \frac{4M}{\delta}\, e^{-R\,\delta/2},$$

so that the result follows immediately.

This theorem may be applied in an obvious way to the bilateral Laplace transform. It is of interest to verify it for Example F of §2. By Stirling's formula (Titchmarsh [1939; p. 259])

$$|\Gamma(\sigma + i\tau)| \sim (2\pi)^{1/2} |\tau|^{(2\sigma-1)/2} e^{-\pi|\tau|/2},$$

$$\left| \cos \frac{\pi}{2}(\sigma + i\tau) \right| \sim \frac{e^{\pi|\tau|/2}}{2}.$$

Hence

$$\Gamma(\sigma + i\tau) \cos \frac{\pi}{2}(\sigma + i\tau) = O(|\tau|^{(2\sigma-1)/2}), \qquad |\tau| \to \infty.$$

But $(2\sigma - 1)/2 < 1$ in $0 < \sigma < 1$, as required.

7. Inversion

The inversion formula for the Laplace–Stieltjes integral is quite analogous to the one developed in Chapter 2 for Dirichlet series. We could prove it in much the same way as before, using Theorem 6 to treat that part of the integral which is extended over $R \leqq t < \infty$, a neighborhood of infinity. To treat the remaining part, over $0 \leqq t \leqq R$, we employ some elementary Fourier analysis. Since this may be used equally well for both parts we abandon the original approach.

Before recalling the Fourier theorem needed we summarize a few standard facts from analysis.

A. $\qquad \int_x^y \frac{\sin t}{t}\, dt < M, \qquad -\infty < x, \qquad y < \infty, \quad$ some $M.$

B. $\qquad f(t) \in C, \quad g(t) \in \uparrow, \quad g(t) \geqq 0, \quad a \leqq t \leqq b,$

$\qquad \Rightarrow \quad \int_a^b f(t)g(t)\, dt = g(b) \int_\xi^b f(t)\, dt, \qquad$ some $\xi, \quad a \leqq \xi \leqq b.$

C. $\qquad f(t) \in L\,(-\infty, \infty) \Rightarrow \lim_{R \to \infty} \int_{-\infty}^{\infty} f(t) \sin Rt\, dt = 0.$

D. $\alpha(t) \in V$ (bounded variation) $a \leqq t \leqq b$; $\alpha(t)$ real; $\alpha(a) = 0$
$v(t) = $ total variation of $\alpha(x)$ on $a \leqq x \leqq t$

$\qquad \Rightarrow \quad \alpha(t) = \alpha_1(t) - \alpha_2(t),$
$\qquad\qquad v(t) = \alpha_1(t) + \alpha_2(t),$

for some nondecreasing functions α_1 and α_2 both vanishing at $a.$

The first of these facts is an immediate consequence of the convergence of the integral

(7.1) $$\int_0^\infty \frac{\sin t}{t}\, dt = \frac{\pi}{2}.$$

Statement B is Bonnet's form of the second law of the mean; statement C is the Riemann–Lebesgue theorem (see Lemma 5.2 of Chapter 4). In the special case in which $\alpha(t)$ is absolutely continuous, statement D follows immediately from the decomposition

$$\alpha(t) = \frac{1}{2} \int_0^t (|\phi(x)| + \phi(x))\, dx - \frac{1}{2} \int_0^t (|\phi(x)| - \phi(x))\, dx = \alpha_1(t) - \alpha_2(t).$$

By use of the above facts we can now prove two lemmas.

Lemma 7.1.

$$1. \quad \alpha(t) \in V \quad \text{(bounded variation)}, \qquad 0 \leq t \leq A$$

$$(7.2) \quad \Rightarrow \qquad \lim_{R \to \infty} \frac{1}{\pi} \int_0^A \alpha(t) \frac{\sin Rt}{t}\, dt = \frac{\alpha(0+)}{2}.$$

By virtue of (7.1) it is no restriction to assume that $\alpha(0+) = 0$. On account of the linearity of the relation (7.2) we may also suppose first that $\alpha(t)$ is real and then (by D) that $\alpha(t) \in \uparrow$. By B

$$I(R) = \int_0^A \alpha(t) \frac{\sin Rt}{t}\, dt$$

$$= \alpha(\delta) \int_\xi^\delta \frac{\sin Rt}{t}\, dt + \int_\delta^A \alpha(t) \frac{\sin Rt}{t}\, dt \qquad 0 \leq \xi \leq \delta,$$

and by A

$$|I(R)| \leq M\alpha(\delta) + \left| \int_\delta^A \frac{\alpha(t)}{t} \sin Rt\, dt \right|.$$

Since $\alpha(t)/t \in L$ on (δ, A) we may apply C to get

$$\overline{\lim_{R \to \infty}} \, |I(R)| \leq M\alpha(\delta) = o(1), \qquad \delta \to 0+.$$

Hence

$$\lim_{R \to \infty} I(R) = 0,$$

as we wished to prove.

The bounded variation condition on $\alpha(t)$ was needed only locally, as the following extension of the previous result shows.

Lemma 7.2.

1. $\alpha(t) \in L, \qquad (-\infty < t < \infty)$;

2. $\alpha(t) \in V$ in a neighborhood of $t = x_0$

$$(7.3) \quad \Rightarrow \quad \lim_{R \to \infty} \frac{1}{\pi} \int_{-\infty}^{\infty} \alpha(t) \frac{\sin R(x_0 - t)}{x_0 - t} \, dt = \frac{\alpha(x_0+) + \alpha(x_0-)}{2}.$$

By hypothesis 2 $\alpha(x_0 + t) \in \dot{V}$ is $-A \leq t \leq A$ for some A. The integral (7.3) is equal to

(7.4)

$$\frac{1}{\pi} \int_0^A [\alpha(x_0 + t) + \alpha(x_0 - t)] \frac{\sin Rt}{t} \, dt + \frac{1}{\pi} \int_{|t| \geq A} \alpha(x_0 + t) \frac{\sin Rt}{t} \, dt.$$

The first of these integrals tends to $[\alpha(x_0 +) + \alpha(x_0 -)]/2$ by Lemma 7.1; the second approaches zero by C since $\alpha(x_0 + t)/t \in L$ on (A, ∞) and on $(-\infty, -A)$. This completes the proof. Condition 1 is practical for our purposes but is stronger than necessary. It could clearly be replaced by $\alpha(t)/(|t| + 1) \in L$ on $(-\infty, \infty)$. For, the Riemann–Lebesgue theorem could still be applied to the second integral (7.4).

Note that Fourier's double integral formula

(7.5)

$$\lim_{R \to \infty} \frac{1}{2\pi} \int_{-R}^{R} \exp(-ix_0 t) \, dt \int_{-\infty}^{\infty} \exp(iyt)\alpha(y) \, dy = \frac{\alpha(x_0+) + \alpha(x_0-)}{2}$$

is an immediate consequence of this lemma.

We now prove the principal inversion formula for the Laplace–Stieltjes integral.

Theorem 7.1.

1. $f(s) = \int_0^{\infty} e^{-st} \, d\alpha(t), \qquad \sigma > \sigma_c$;

2. $a > 0, \qquad a > \sigma_c$

$$(7.6) \quad \Rightarrow \quad \lim_{R \to \infty} \frac{1}{2\pi i} \int_{a-iR}^{a+iR} f(s) \frac{e^{st}}{s} \, ds = \frac{\alpha(t+) + \alpha(t-)}{2}, \qquad t > 0$$

$$= \frac{\alpha(0+)}{2}, \qquad t = 0$$

$$= 0, \qquad t < 0.$$

By Theorem 5.2

$$(7.7) \qquad \frac{f(s)}{s} = \int_0^\infty e^{-sy} \alpha(y) \, dy,$$

the integral converging absolutely on the line $\sigma = a$ and uniformly (by Theorem 4.1) on any finite segment of it. Hence the integration with respect to s described in (7.6) may be performed *under* the integral sign (7.7) to produce

$$\frac{1}{2\pi i} \int_{a-iR}^{a+iR} f(s) \frac{e^{st}}{s} \, ds = \frac{1}{2\pi i} \int_0^\infty \alpha(y) \, dy \int_{a-iR}^{a+iR} e^{s(t-y)} \, ds$$

$$(7.8) \qquad\qquad = \frac{1}{\pi} \int_0^\infty \alpha(y) e^{a(t-y)} \frac{\sin R(t-y)}{t-y} \, dy.$$

Defining $\alpha(y)$ to be 0 for $y < 0$, $\alpha(y)e^{-ay} \in L$ on $(-\infty, \infty)$ and Lemma 7.2 is applicable to the integral (7.8), giving the desired result.

A uniqueness theorem for the Laplace–Stieltjes integral is now immediate. Of course, the determining function can be altered at many points without changing the generating function; so some sort of normalization is needed. We say that a function of bounded variation $\alpha(t)$ is *normalized*, $\alpha \in V^*$, on $0 \leq t < R$ if and only if

$$\alpha(0) = 0, \qquad \alpha(t) = \frac{\alpha(t+) + \alpha(t-)}{2}, \qquad 0 < t < R.$$

We can then prove the following uniqueness result.

Theorem 7.2.

$$1. \quad \alpha(t) \in V^*, \qquad 0 \leq t < R, \quad \text{every } R > 0;$$

$$(7.9) \qquad 2. \quad \int_0^\infty e^{-st} \, d\alpha(t) \equiv 0, \qquad \sigma > \sigma_0, \quad \text{some } \sigma_0$$

$$\Rightarrow \qquad \alpha(t) \equiv 0, \qquad 0 \leq t < \infty.$$

For, we have only to choose the constant a of Theorem 7.1 > 0 and $> \sigma_0$ to see that the limit (7.6) is zero for $t > 0$. But the right-hand side of (7.6) is $\alpha(t)$, and the proof is complete.

Corollary 7.2.

$$\text{(7.10)} \quad \begin{aligned} &1. \quad \phi(t) \in L, \quad 0 \leqq t < R, \quad \text{every } R > 0; \\ &2. \quad \int_0^\infty e^{-st}\phi(t)\, dt \equiv 0, \quad \sigma > \sigma_0, \quad \text{some } \sigma_0 \end{aligned}$$

$$\Rightarrow \quad \phi(t) = 0 \quad \text{for almost all } t > 0.$$

For, equations (7.9) and (7.10) are the same if

$$\text{(7.11)} \qquad \alpha(t) = \int_0^t \varphi(y)\, dy.$$

The present $\alpha(t)$ is continuous and hence normalized, and by the theorem is identically zero. Hence its derivative, which is almost everywhere equal to ϕ, is identically zero.

This corollary is of course in constant use when problems are solved by use of tables of Laplace transforms. See Theorem 1 of Chapter 1.

Let us invert next the Laplace–Lebesgue transform. Here the bilateral case can be treated as simply as the unilateral one. Note that formal differentiation of equation (7.6) would cancel the factor $(1/s)$ from the integrand and would give $\phi(t)$ in (7.11). The result is true under limited circumstances.

Theorem 7.3.

$$\text{(7.12) 1.} \quad f(s) = \int_{-\infty}^\infty e^{-st}\phi(t)\, dt \quad \text{converges absolutely for } \sigma = a;$$

$$2. \quad \phi(t) \in V \quad \text{in a neighborhood of } t_0$$

$$\Rightarrow \quad \lim_{R \to \infty} \frac{1}{2\pi i} \int_{a-iR}^{a+iR} f(s) \exp st_0 \, ds = \frac{\phi(t_0+) + \phi(t_0-)}{2}.$$

For, the integral (7.12) converges uniformly on the line segment $s = a + iy$, $R \leqq y \leqq R$. Hence integration under the integral sign over that range is permitted:

$$\frac{1}{2\pi i}\int_{a-iR}^{a+iR} f(s)\exp st_0\,ds = \frac{1}{2\pi i}\int_{-\infty}^{\infty}\phi(u)\,du\int_{a-iR}^{a+iR}\exp s(t_0-u)\,ds$$

$$= \frac{1}{\pi}\int_{-\infty}^{\infty}\phi(u)\exp a(t_0-u)\,\frac{\sin R(t_0-u)}{t_0-u}\,du.$$

But $\phi(u)e^{-au}\in L\;(-\infty<u<\infty)$ by hypothesis 1; so that we are in a position to apply Lemma 7.2. This gives the stated result. Contrast Theorems 7.1 and 7.3. In the former the constant a must be >0, not in the latter; in the former absolute convergence is not in the hypothesis but is there in the latter.

Corollary 7.3a.

1. $f(s)=\displaystyle\int_0^{\infty} x^{s-1}\phi(x)\,dx$ converges absolutely for $\sigma=a$;

2. $\phi(x)\in V$ in a neighborhood of x_0

$$\Rightarrow\qquad \lim_{R\to\infty}\frac{1}{2\pi i}\int_{a-iR}^{a+iR} f(s)x_0^{-s}\,ds = \frac{\phi(x_0+)+\phi(x_0-)}{2}.$$

This is just a restatement of the theorem in the Mellin form.

Corollary 7.3b.

1. $\phi(t)\in L,\quad(-\infty<t<\infty)$;

2. $\phi(t)\in V$ in a neighborhood of t_0;

(7.13)   3. $g(y)=\displaystyle\int_{-\infty}^{\infty} e^{-iyt}\phi(t)\,d\bar{t}$

$$\Rightarrow\quad \lim\frac{1}{2\pi}\int_{-R}^{R} g(y)\exp iyt_0\,dy = \frac{\phi(t_0+)+\phi(t_0-)}{2}.$$

This is the special case $a=0$ of the theorem in which we have set $g(y)=f(iy)$. It gives an inversion of the Fourier transform (7.13).

Assuming the validity of transform G of §2 let us invert it as an example:

$$(7.14)\quad \exp s^2 c = \int_{-\infty}^{\infty} e^{-st}\,\frac{\exp\!\left[-\dfrac{t^2}{(4c)}\right]}{(4\pi c)^{1/2}}\,dt,\qquad -\infty<\sigma<\infty,\quad c>0.$$

By Theorem 7.3 with $a = 0$

$$(7.15) \qquad \frac{\exp\left[-\dfrac{t^2}{(4c)}\right]}{(4\pi c)^{1/2}} = \frac{1}{2\pi} \int_{-\infty}^{\infty} \exp(-y^2 c) e^{iyt}\, dy.$$

Note that there is no need to use the Cauchy principle value of the integral since it clearly converges. But we can check (7.15) in this special case by (7.14) itself by setting $s = -it/(2c)$ therein (renaming the variable of integration). In particular, if $c = \frac{1}{2}$ we obtain the classic formula

$$\exp\left(-\frac{t^2}{2}\right) = \frac{1}{(2\pi)^{1/2}} \int_{-\infty}^{\infty} e^{iyt} \exp\left(-\frac{y^2}{2}\right) dy, \qquad -\infty < t < \infty,$$

which states that $\exp(-t^2/2)$ is its own Fourier transform except for a constant factor.

8. Convolutions

When two power series, $\sum a_k z^k$ and $\sum b_k z^k$, are multiplied together the result is a new power series $\sum c_k z^k$ with

$$c_n = \sum_{k=0}^{n} a_k b_{n-k}, \qquad n = 0, 1, 2, \ldots.$$

A similar result holds for Laurent series, but now

$$c_n = \sum_{k=-\infty}^{\infty} a_k b_{n-k}, \qquad n = 0, \pm 1, \pm 2, \ldots.$$

Following our usual analogy with power series we would expect that the product of two Laplace integrals, $\int e^{-st} a(t)\, dt$, $\int e^{-st} b(t)\, dt$ is the Laplace integral $\int e^{-st} c(t)\, dt$, where

$$c(x) = \int_0^x a(t)b(x - t)\, dt, \qquad 0 \le x < \infty,$$

$$c(x) = \int_{-\infty}^{\infty} a(t)b(x - t)\, dt, \qquad -\infty < x < \infty,$$

in the unilateral and bilateral cases, respectively. We show here that this is the case under suitable restrictions on the functions involved.

We introduce the following definition and notation.

Definition 8. The convolution of $\phi(x)$ with $\psi(x)$ is

$$(8.1) \qquad \omega(x) = \varphi(x) * \psi(x) = \int_{-\infty}^{\infty} \phi(t)\psi(x - t) \, dt$$

when this integral converges.

A simple change of variable shows that $\phi * \psi = \psi * \phi$. In particular, if ϕ and ψ vanish for $t < 0$, (8.1) reduces to

$$(8.2) \qquad \omega(x) = \int_{0}^{x} \phi(t)\psi(x - t) \, dt,$$

so that the unilateral case is a special case of the bilateral. It will result from the following theorem that $\omega(x)$ is defined almost everywhere when ϕ and $\psi \in L\,(-\infty, \infty)$.

Theorem 8.

$$1. \quad f(s) = \int_{-\infty}^{\infty} e^{-st}\phi(t) \, dt \quad \text{converges absolutely for } \sigma = a;$$

$$2. \quad g(s) = \int_{-\infty}^{\infty} e^{-st}\psi(t) \, dt \quad \text{converges absolutely for } \sigma = a$$

$$\Rightarrow \text{A.} \quad \omega(x) = \int_{-\infty}^{\infty} \phi(t)\psi(x - t) \, dt \quad \begin{array}{l}\text{converges absolutely almost}\\ \text{everywhere;}\end{array}$$

$$(8.3) \quad \text{B.} \quad f(s)g(s) = \int_{-\infty}^{\infty} e^{-st}\omega(t) \, dt \quad \text{converges absolutely for } \sigma = a.$$

For,

$$(8.4) \qquad \int_{-\infty}^{\infty} |\phi(t)| \, dt \int_{-\infty}^{\infty} e^{-ay}|\psi(y - t)| \, dy < \infty,$$

since the inner integral is equal to

$$e^{-at} \int_{-\infty}^{\infty} e^{-ay}|\psi(y)| \, dy$$

and this function of t when multiplied by $\phi(t) \in L$ by hypotheses 1 and 2. Hence by Fubini's theorem we may invert the order of integration in (8.4) to obtain

$$(8.5) \qquad \int_{-\infty}^{\infty} e^{-ay} \, dy \int_{-\infty}^{\infty} |\phi(t)\psi(y - t)| \, dt < \infty.$$

As a consequence, the integral defining $\omega(x)$ converges absolutely almost everywhere and the absolute convergence of the integral (8.3) is proved. Finally, equation (8.3) is established by inverting the order of integration in

$$\int_{-\infty}^{\infty} \phi(t)\, dt \int_{-\infty}^{\infty} e^{-sy}\psi(y-t)\, dy = f(s)g(s), \qquad \sigma = a.$$

This concludes the proof. If $a = 0$, then ϕ and $\psi \in L$ and (8.5) shows that $\omega(x) \in L$ also.

Corollary 8.

1. $f(s) = \displaystyle\int_{0}^{\infty} x^{s-1}\phi(x)\, dx$ converges absolutely for $\sigma = a$;

2. $g(s) = \displaystyle\int_{0}^{\infty} x^{s-1}\psi(x)\, dx$ converges absolutely for $\sigma = a$

$\Rightarrow$ A. $\omega(x) = \displaystyle\int_{0}^{\infty} \phi(t)\psi\left(\frac{x}{t}\right)\frac{dt}{t}$ converges absolutely, almost all
$$x > 0;$$

B. $f(s)g(s) = \displaystyle\int_{0}^{\infty} x^{s-1}\omega(x)\, dx$ converges absolutely for $\sigma = a$.

This is the Mellin form of the "product theorem." It follows by an easy change of variable. Note the factor $(1/t)$.

As an application of the product theorem let us reprove Lemma 7.2 of Chapter 3. From equation (2.1) of that chapter

$$\zeta(s) = s\int_{1}^{\infty} \frac{[x]}{x^{s+1}}\, dx = s\int_{0}^{1} x^{s-1}\left[\frac{1}{x}\right] dx, \qquad \sigma > 1.$$

Hence

$$\zeta(s)\Gamma(s) = \int_{0}^{1} x^{s-1}\left[\frac{1}{x}\right] dx \int_{0}^{\infty} x^{s-1}xe^{-x}\, dx.$$

By Corollary 8

$$\zeta(s)\Gamma(s) = \int_{0}^{\infty} x^{s-1}\omega(x)\, dx, \qquad \sigma > 1,$$

where

$$(8.6) \qquad \omega(x) = \int_0^1 \left[\frac{1}{t}\right] \frac{x}{t} e^{-x/t} \frac{dt}{t} = x \int_1^\infty [t] e^{-xt}\, dt, \qquad \sigma > 1.$$

Replacing the lower limit 1 in the second integral (8.6) by $\frac{1}{2}$ (leaving its value unchanged), we obtain

$$\omega(x) = \int_{1/2}^\infty e^{-xt}\, d[t] = \sum_{k=1}^\infty e^{-kx} = \frac{1}{e^x - 1}, \qquad x > 0,$$

$$\zeta(s)\Gamma(s) = \int_0^\infty \frac{x^{s-1}}{e^x - 1}\, dx, \qquad \sigma > 1.$$

9. Fractional Integrals

We shall seek to obtain here the analog of Theorem 9.2 of Chapter 2. The number p of that theorem was a positive integer. Here we shall permit it to be any positive number. First we introduce the notion of fractional integrals. Recall that the nth iterated integral of a function $\alpha(t)$ may be put in the form

$$\frac{1}{(n-1)!} \int_0^x (x-t)^{n-1} \alpha(t)\, dt \qquad n = 1, 2, 3, \ldots.$$

The nth derivative with respect to x of this function is $\alpha(x)$. With this in mind the following definition is a natural one.

Definition 9. The fractional integral of $\alpha(x)$ of order ρ is

$$(9.1) \qquad \alpha_\rho(x) = D^{-\rho}\alpha(x) = \frac{1}{\Gamma(\rho)} \int_0^x (x-t)^{\rho-1}\alpha(t)\, dt, \qquad \rho > 0.$$

The fractional derivative of $\alpha(x)$ of order ρ is

$$(9.2) \qquad \alpha^{(\rho)}(x) = D^\rho \alpha(x) = D^{[\rho]+1}(D^{\rho-[\rho]-1}\alpha(x)), \qquad \rho < 0.$$

The first is called the fractional integral of order ρ; the second is the fractional derivative of order ρ. $[\rho]$ is the greatest integer $\leqq \rho$.

Note that $\rho - [\rho] - 1 < 0$ so that the function in parenthesis (9.2) is defined by (9.1). For example,

$$D^{-\rho}x^r = \frac{1}{\Gamma(\rho)} \int_0^x (x - t)^{\rho-1} t^r \, dt = \frac{x^{\rho+r}\Gamma(r + 1)\Gamma(\rho)}{\Gamma(\rho)\Gamma(r + \rho + 1)}, \qquad \rho > 0, \quad r > -1,$$

and if $\rho = n$, a positive integer, this reduces to

$$\frac{x^{r+n}}{(r + n)(r + n - 1) \cdots (r + 1)}.$$

As another example

$$\sqrt{D}\, X^r = \frac{\Gamma(r + 1)}{\Gamma(r + \tfrac{1}{2})} x^{r-1/2}, \qquad r > -\tfrac{1}{2}.$$

Note that if $\alpha(0) = 0$, then

$$\alpha_\rho(x) = \frac{1}{\Gamma(\rho + 1)} \int_0^x (x - t)^\rho \, d\alpha(t).$$

To obtain the desired inversion formula we need a preliminary result.

Lemma 9.

$$1. \quad \alpha(x) \in V, \qquad 0 \leqq x < \infty;$$

$$2. \quad \rho > 0$$

$$\Rightarrow \qquad \alpha_\rho(x) \in V \cdot C, \qquad 0 \leqq x < \infty.$$

Since the operation $D^{-\rho}$ is clearly a linear one it is sufficient to assume that $\alpha(t)$ is real, $\alpha(0) = 0$, $\alpha(t) \in \uparrow$, and to prove that $\alpha_\rho(x) \in \uparrow \cdot C$. For $\delta > 0$, $x \geqq 0$,

$$\Gamma(\rho + 1)[\alpha_\rho(x + \delta) - \alpha_\rho(x)]$$

$$= \int_x^{x+\delta} (x + \delta - t)^\rho \, d\alpha(t) + \int_0^x [(x + \delta - t)^\rho - (x - t)^\rho] \, d\alpha(t).$$

The right-hand side of this equation is $\geqq 0$ since the two integrands are $\geqq 0$ and $\alpha(t) \in \uparrow$. Also both integrands are decreasing if $\rho > 1$ though the second is increasing if $0 < \rho < 1$. Hence

(9.3) $\Gamma(\rho + 1)[\alpha_\rho(x + \delta) - \alpha_\rho(x)]$

$$\leqq \delta^\rho \int_x^{x+\delta} d\alpha(t) + [(x + \delta)^\rho - x^\rho] \int_0^x d\alpha(t), \qquad \rho > 1$$

$$\leqq \delta^\rho \int_x^{x+\delta} d\alpha(t) + \delta^\rho \int_0^x d\alpha(t)$$

$$= \delta^\rho \int_0^{x+\delta} d\alpha(t), \qquad 0 < \rho \leqq 1.$$

In either case the right-hand side tends to zero with δ so that $\alpha_\rho(x +) = \alpha_\rho(x)$. In like manner we have for $0 < \delta < x$

(9.4) $\Gamma(\rho + 1)[\alpha_\rho(x) - \alpha_\rho(x - \delta)]$

$$\leqq \delta^\rho \int_{x-\delta}^x d\alpha(t) + [x^\rho - (x - \delta)^\rho] \int_0^{x-\delta} d\alpha(t), \qquad \rho > 1,$$

$$\leqq \delta^\rho \int_{x-\delta}^x d\alpha(t) + \delta^\rho \int_0^{x-\delta} d\alpha(t)$$

$$= \delta^\rho \int_0^x d\alpha(t), \qquad 0 < \rho \leqq 1.$$

This shows that $\alpha_\rho(x -) = \alpha_\rho(x)$ for $x > 0$ so that $\alpha_\rho(x)$ is continuous, as stated.

We can now prove the desired result.

Theorem 9.

$$1. \quad f(s) = \int_0^\infty e^{-st}\, d\alpha(t), \qquad \sigma > \sigma_c;$$

$$2. \quad a > 0, \quad a > \sigma_c, \quad \rho > 0$$

$$\Rightarrow \qquad \lim_{R \to \infty} \frac{1}{2\pi i} \int_{a-iR}^{a+iR} \frac{f(s)e^{st}}{s^{\rho+1}}\, ds = \alpha_\rho(t), \qquad 0 < t < \infty.$$

By Theorem 5.2

(9.5) $$\frac{f(s)}{s} = \int_0^\infty e^{-st}\alpha(t)\, dt, \qquad \sigma > \sigma_c, \quad \sigma > 0,$$

the integral converging absolutely. Now use Theorem 8.2, as applied to unilateral integrals, to multiply the integral (9.5) by the integral

$$\frac{1}{s^\rho} = \int_0^\infty e^{-st} \frac{t^{\rho-1}}{\Gamma(\rho)} \, dt, \qquad \sigma > 0.$$

Since $\alpha_\rho(t)$ is precisely the convolution obtained from that theorem we obtain

$$\frac{f(s)}{s^{\rho+1}} = \int_0^\infty e^{-st} \alpha_\rho(t) \, dt, \qquad \sigma > 0, \quad \sigma > \sigma_c.$$

We may now invert this integral by use of Theorem 7.3, noting that the hypotheses thereof are here satisfied by virtue of Lemma 9.

Note that a Dirichlet series,

$$f(s) = \sum_{k=1}^\infty a_k \exp(-\lambda_k s) = \int_0^\infty e^{-st} \, d\alpha(t),$$

is included in the theorem. In this case

$$\alpha_\rho(x) = \int_0^x \frac{(x-t)^\rho}{\Gamma(\rho+1)} \, d\alpha(t) = \sum_{\lambda_k \leqq x} \frac{(x-\lambda_k)^\rho}{\Gamma(\rho+1)} a_k.$$

We thus recapture Theorem 9.2, Chapter 2, if ρ is integral. In particular, for the Riemann zeta-function we have

$$\frac{1}{2\pi i} \int_{a-i\infty}^{a+i\infty} \frac{\zeta'(s)}{\zeta(s)} \frac{x^s}{s^2} \, ds = -\int_1^x \frac{\psi(t)}{t} \, dt, \qquad a > 1,$$

$$\frac{1}{2\pi i} \int_{a-i\infty}^{a+i\infty} \frac{\zeta'(s)}{\zeta(s)} \frac{x^s}{s^3} \, ds = -\int_1^x \frac{dt}{t} \int_1^y \frac{\psi(y)}{y} \, dy, \qquad a > 1.$$

These are two formulas used in Chapter 4 but there obtained from the theory of Dirichlet series. No Cauchy value is needed for these complex integrals. They converge since [(8.2), Chapter 3]

$$\frac{\zeta'(a+iy)}{\zeta(a+iy)} = O(\log^9 |y|), \qquad |y| \to \infty.$$

10. Analytic Behavior of Generating Functions

We prove here a result which generalizes Theorem 11.1 of Chapter 2. A generating function need have no singularity on the axis of convergence. However, if the determining function is nondecreasing the following theorem shows that there is surely a singularity on the axis.

Theorem 10.1.

$$1. \quad f(s) = \int_0^\infty e^{-st}\, d\alpha(t), \qquad \sigma > \sigma_c > -\infty;$$

$$2. \quad \alpha(t) \in \uparrow, \qquad 0 < t < \infty$$

$$\Rightarrow \qquad f(s) \notin A \quad \text{at} \quad s = \sigma_c.$$

As in Theorem 11.1, Chapter 2, we may assume $\sigma_c = 0$. We shall deduce a contradiction from the assumption $f(s) \in A$ at $s = 0$. This would imply $f(s) \in A$ in some neighborhood of $s = 0$ and hence that the Taylor expansion

$$(10.1) \qquad f(s) = \sum_{k=0}^\infty f^{(k)}(1) \frac{(s-1)^k}{k!}$$

would converge for some real and negative value of s, $s = -\delta$. But by Theorem 4.2 equation (10.1) becomes for $s = -\delta$

$$f(-\delta) = \sum_{k=0}^\infty \frac{(\delta+1)^k}{k!} \int_0^\infty e^{-t} t^k \, d\alpha(t) < \infty.$$

And now since all factors are positive and $\alpha(t) \in \uparrow$ we may interchange the symbols $\sum, \int$ (using the series analog of Fubini's theorem) to obtain

$$\int_0^\infty e^{-t} \sum_{k=0}^\infty \frac{(\delta+1)^k t^k}{k!} \, d\alpha(t) = \int_0^\infty e^{\delta t} \, d\alpha(t) < \infty.$$

But this contradicts the hypotheses $\sigma_c = 0$.

As an example, consider formula 6 of §2, Chapter 1,

$$\frac{2s}{1-s^2} = \int_0^\infty e^{-st} \cosh t \, dt.$$

Here $\alpha(t) = \sinh t \in \uparrow$, $\sigma_c = 1$, and the left-hand side is indeed singular at $s = 1$, as predicted.

Let us prove here a simple result, frequently useful. It is that the asymptotic behavior of the generating function at ∞ is determined by the behavior of the determining function near the origin.

Theorem 10.2.

$$(10.2) \qquad 1. \quad f(s) = \int_0^\infty e^{-st}\, d\alpha(t), \qquad \sigma > \sigma_c$$

$$\Rightarrow \qquad f(+\infty) = \lim_{\sigma \to +\infty} f(\sigma) = \alpha(0+).$$

By Theorem 5.2

$$f(\sigma) = \sigma \int_0^\infty e^{-\sigma t}\alpha(t)\,dt, \qquad \sigma > \sigma_c, \quad \sigma > 0.$$

Since

$$\alpha(0+) = \sigma \int_0^\infty e^{-\sigma t}\alpha(0+)\,dt,$$

we may assume, without loss of generality, that $\alpha(0+) = 0$. Then for any $\delta > 0$ and any positive $\sigma_0 > \sigma_c$

$$f(\sigma) = \sigma \int_0^\delta e^{-\sigma t}\alpha(t)\,dt + \sigma \int_\delta^\infty \exp[-(\sigma - \sigma_0)t]\exp(-\sigma_0 t)\alpha(t)\,dt,$$

$$|f(\sigma)| \leq \sup_{0 \leq t \leq \delta} |\alpha(t)|\,\sigma \int_0^\infty e^{-\sigma t}\,dt$$

$$+ \sigma \exp[-(\sigma - \sigma_0)\delta]\int_\delta^\infty \exp(-\sigma_0 t)|\alpha(t)|\,dt,$$

$$\varlimsup_{\sigma \to \infty} |f(\sigma)| \leq \sup_{0 \leq t \leq \delta} |\alpha(t)|.$$

But by our assumption the right-hand side tends to zero with δ. Since the left-hand side is independent of δ it must be zero.

One useful consequence is that a Laplace–Lebesgue transform always approaches zero as $\sigma \to +\infty$:

$$\lim_{\sigma \to \infty} \int_0^\infty e^{-\sigma t}\phi(t)\,dt = 0.$$

Note also that if $\alpha(\infty)$ exists in (10.2) then $f(0+) = \alpha(\infty)$. This is an immediate consequence of Theorem 4.1.

11. Representation

We shall obtain here two sets of sufficient conditions that are highly practical for the representation of functions as bilateral and as unilateral Laplace integrals. Unless one restricts the determining function to lie in some special class ($\uparrow$, B, L^p, etc.) it is impractical to ask for necessary and sufficient conditions. In §12 we shall prove such a theorem and in Chapter 6 another, Theorem 7.

A result for the bilateral transform is usually attributed to H. Hamburger [1921, p. 243].

Theorem 11.1.

1. $f(s) \in A,\qquad \alpha < \sigma < \beta$;

2. $f(\sigma_0 + iy) \in L\quad (-\infty < y < \infty)\quad$ for each $\sigma_0,\quad \alpha < \sigma_0 < \beta$;

3. $\quad \lim f(\sigma + iy) = 0\quad$ uniformly $\quad \alpha < \sigma < \beta$

$$(11.1)\quad \Rightarrow \qquad f(s) = \int_{-\infty}^{\infty} e^{-st}\phi(t)\, dt$$

for some function $\phi(t)$, the integral converging absolutely in $\alpha < \sigma < \beta$.

By Fourier's integral formula, (7.5), applicable by virtue of hypothesis 2, we have

$$f(\sigma_0 + iy) = \lim_{R\to\infty} \frac{1}{2\pi} \int_{-R}^{R} e^{-iyt}\, dt \int_{-\infty}^{\infty} e^{irt} f(\sigma_0 + ir)\, dr$$

$$(11.2) \qquad = \frac{1}{2\pi} \int_{-\infty}^{\infty} \exp[-(\sigma_0 + iy)t]\, dt$$

$$\times \int_{-\infty}^{\infty} \exp[(\sigma_0 + ir)t] f(\sigma_0 + ir)\, dr,$$

provided that the outer integral (11.2) converges. We shall show that it does. Set

$$(11.3) \qquad \phi(t) = \frac{1}{2\pi} \int_{-\infty}^{\infty} e^{(\sigma + ir)t} f(\sigma + ir)\, dr.$$

We note first that this integral is indeed independent of σ, as the notation suggests. By Cauchy's theorem, valid in the presence of hypothesis 1, this will be true if

$$\lim_{|r|\to\infty} \int_{\sigma_1}^{\sigma_2} e^{(\sigma + ir)t} f(\sigma + ir)\, d\sigma = 0$$

for any pair of numbers σ_1 and σ_2 in $\alpha < \sigma < \beta$. But this clearly follows from hypothesis 3. From (11.3) and hypothesis 2

$$|\phi(t)| \leqq M_\sigma e^{\sigma t} = \frac{e^{\sigma t}}{2\pi} \int_{-\infty}^{\infty} |f(\sigma + ir)| \, dr.$$

If in this estimate we choose σ near α we can use it to show that

$$\int_0^\infty e^{-st}\phi(t) \, dt$$

converges absolutely for $\sigma > \alpha$; if we choose σ near β we see that

$$\int_{-\infty}^0 e^{-st}\phi(t) \, dt$$

converges absolutely for $\sigma < \beta$. Thus the absolute convergence of (11.1) is established, and we were correct in dispensing with the Cauchy value in (11.2).

As an illustration consider the transform C of §2. The generating function $\sec(\pi s/2)$ is analytic in the strip $-1 < \sigma < 1$ and

$$\left| \frac{\sec \pi(\sigma + iy)}{2} \right|^2 \leqq \frac{1}{\cosh 2\pi y - 1} = O(e^{-2\pi|y|}), \qquad |y| \to \infty.$$

Hence hypotheses 2 and 3 are satisfied in any vertical strip. Thus the region of validity of formula C should be $|\sigma| < 1$, as indicated in §2.

The corresponding result for the unilateral transform is contained in the following theorem.

Theorem 11.2.

1. $f(s) \in A, \qquad \alpha < \sigma < \infty$;

2. $f(\sigma_0 + iy) \in L \quad (-\infty < y < \infty)$ for each $\sigma_0, \qquad \alpha < \sigma_0 < \infty$;

3. $\lim_{|s| \to \infty} f(s) = 0$

$$\Rightarrow \qquad f(s) = \int_0^\infty e^{-st}\phi(t) \, dt$$

for some function $\phi(t)$, the integral converging absolutely in $\alpha < \sigma < \infty$.

Since the inequality

$$|f(\sigma + iy| < \varepsilon$$

for $|\sigma + iy| > R$ and $\sigma > \alpha$ certainly implies the same inequality for $|y| > R$ and $\sigma > \alpha$, it is clear that hypothesis 3 of the present theorem implies hypothesis 3 of Theorem 11.1 for any $\beta > \alpha$. Hence the conclusion of the latter theorem must hold:

$$f(s) = \int_{-\infty}^{\infty} e^{-st}\phi(t)\, dt,$$

the integral converging absolutely for $\alpha < \sigma < \infty$. [$\phi(t)$ cannot depend on β since it is uniquely determined by formula (11.3)]. It remains only to show that $\phi(t)$ vanishes for negative t.

By Cauchy's theorem the integral of $f(s)e^{st}$ over the line segment $\sigma = c > \alpha$, $-R \leqq y \leqq R$, is equal to the integral of that function over the semicircumference $s = c + Re^{i\theta}$, $-\dfrac{\pi}{2} \leqq \theta \leqq \dfrac{\pi}{2}$:

(11.4) $\quad i\displaystyle\int_{-R}^{R} e^{(c+iy)t}f(c + iy)\, dy$

$$= iR \int_{-\pi/2}^{\pi/2} \exp[t(c + Re^{i\theta})]f(c + Re^{i\theta})e^{i\theta}\, d\theta.$$

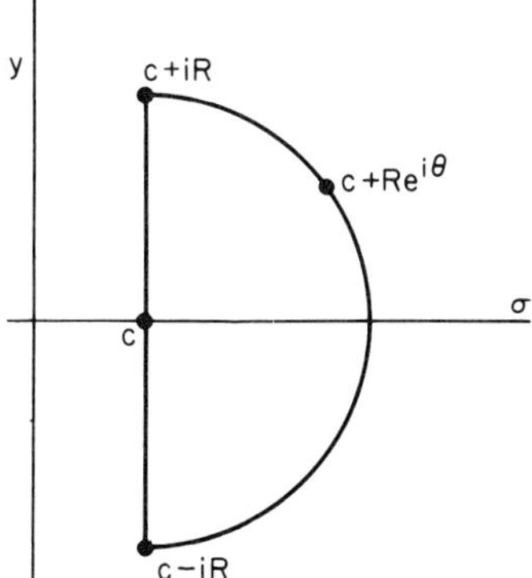

The integral on the right is in absolute value at most equal to

(11.5) $\qquad Re^{tc} \displaystyle\sup_{-\pi/2 \leqq \theta \leqq \pi/2} |f(c + Re^{i\theta})| \int_{-\pi/2}^{\pi/2} e^{tR\cos\theta}\, d\theta.$

But for $t < 0$

$$R \int_{-\pi/2}^{\pi/2} e^{tR\cos\theta}\, d\theta \leqq 2R \int_{0}^{\pi/2} \exp\left[tR\left(1 - \frac{2\theta}{\pi}\right)\right] d\theta < \frac{\pi}{t}.$$

By hypothesis 3 it is thus clear that the function (11.5) tends to zero when $R \to \infty$. The same must be true of the first integral (11.4). Hence by formula (11.3) $\phi(t) = 0$ for $t < 0$.

As examples consider the two function $1/(s^2 + 1)$ and $1/(s^2 - 1)$. Both satisfy condition 3 for any α and both belong to L on any vertical line not passing through a pole. Hence σ_c in these cases must be governed by hypothesis 1, so that $\sigma_c = 0$ for the first and $\sigma_c = 1$ for the second:

$$\frac{1}{s^2 + 1} = \int_0^\infty e^{-st} \sin t \, dt, \qquad \sigma > 0,$$

$$\frac{1}{s^2 - 1} = \int_0^\infty e^{-st} \sinh t \, dt, \qquad \sigma > 1.$$

12. Generating Functions Analytic at Infinity

In this section we shall restrict the determining function to be entire. In this way we shall be able to arrive at a necessary and sufficient condition for representation. Let us first recall certain facts about entire functions and introduce a notation.

The order of an entire function is a real number which describes the rate of growth of its modulus as the variable approaches ∞, as compared with the same rate for exponential functions. Thus the order is ρ for $f(z)$ if

$$f(z) = O\,(\exp |z|^{\rho + \varepsilon}), \qquad |z| \to \infty$$

for every $\varepsilon > 0$ and for no $\varepsilon < 0$. For example $\exp(z^n)$ is entire of order n, $n = 0, 1, 2, \ldots$; $\cos \sqrt{z}$ is entire of order $1/2$. For $\exp\{e^z\}$, $\rho = \infty$.

If $0 < \rho < \infty$ the type of $f(z)$ is γ if

$$(12.1) \qquad f(z) = O\,(\exp(\gamma + \varepsilon)|z|^\rho), \qquad |z| \to \infty$$

for every $\varepsilon > 0$ and for no $\varepsilon < 0$. For example $z^2 e^{2z}$ is of order 1 and of type 2; $\cos a\sqrt{z}$ is of order $1/2$ and of type $|a|$.

Let us now introduce the notation $f(z) \in \{\rho, \gamma\}$ by the following definition. (See, for example, R. P. Boas [1954, p. 18].)

Definition 12. The entire function $f(z) \in \{\rho, \gamma\}$, or has growth $\{\rho, \gamma\}$

$\Leftrightarrow$

    A. $f(z)$ is of order $< \rho$,   or

    B. $f(z)$ is of order ρ and of type $\leq \gamma$.

Part of the usefulness of this notation stems from the fact that the class $\{\rho_1, \gamma_1\}$ is included in $\{\rho_2, \gamma_2\}$ if only $\rho_1 < \rho_2$. Thus $\sin 3z \in \{1, 3\}$, but also $\sin 3z \in \{2, \gamma\}$ for any γ. If a power series expansion for $f(z)$ is available there is a convenient formula for determining its growth (R. P. Boas [1954, p. 11]).

Lemma 12.

$$\sum_{k=0}^{\infty} b_k z^k \in \{\rho, \gamma\} \Leftrightarrow \varlimsup_{k \to \infty} k |b_k|^{\rho/k} \leq \rho \gamma e,$$

$$\sum_{k=0}^{\infty} \frac{a_k}{k!} z^k \in \{\rho, \gamma\} \Leftrightarrow \varlimsup_{k \to \infty} \left(\frac{k}{e}\right)^{1-\rho} |a_k|^{\rho/k} \leq \rho \gamma.$$

These two formulas are seen to be equivalent by use of Stirling's formula. A convenient way to remember them is to recall the expansion

$$\exp \gamma z^n = \sum_{k=0}^{\infty} \frac{\gamma^k}{k!} z^{nk} \in \{n, \gamma\}.$$

Since for $n > 1$ many powers of z are missing in the expansion

$$\lim_{k \to \infty} k |b_k|^{n/k} = \lim_{k \to \infty} nk |b_{nk}|^{1/k}$$

$$= \lim_{k \to \infty} \frac{nk\gamma}{ke^{-1}(2\pi k)^{1/(2k)}} = n\gamma e.$$

We can now obtain the desired representation criterion.

Theorem 12.

(12.2) $$f(s) = \int_0^{\infty} e^{-st} \phi(t)\, dt, \qquad \phi(t) \in \{1, \gamma\}$$

$$\Leftrightarrow \qquad f(s) \in A \text{ for } |s| > \gamma \text{ and } f(\infty) = 0.$$

Assume first the representation (12.2). Since

$$\phi(t) = O\left(\exp(\gamma + \varepsilon)t\right), \qquad t \to \infty,$$

by (12.1), the integral (12.2) converges absolutely for $\sigma > \gamma + \varepsilon$, every $\varepsilon > 0$, and hence for $\sigma > \gamma$. If

$$\phi(t) = \sum_{k=0}^{\infty} \frac{a_k \, t^k}{k!},$$

then by Lemma 12

$$(12.3) \qquad \overline{\lim_{k \to \infty}} \; |a_k|^{1/k} \leqq \gamma.$$

But

$$(12.4) \qquad f(s) = \int_0^{\infty} e^{-st} \sum_{k=0}^{\infty} \frac{a_k \, t^k}{k!} \, dt = \sum_{k=0}^{\infty} \frac{a_k}{s^{k+1}}$$

provided that

$$(12.5) \qquad \sum_{k=0}^{\infty} \frac{|a_k|}{\sigma^{k+1}} < \infty.$$

Here we have used the series analog of Fubini's theorem to integrate term by term in (12.4). Inequality (12.5) follows for $\sigma > \gamma$ from (12.3) by the familiar formula for the radius of convergence of a power series. Hence equation (12.4) is valid in the half-plane $\sigma > \gamma$. But since series (12.4) converges for $|s| > \gamma$ it serves to extend $f(s)$ analytically into that region and the desired conclusion follows. It is clear from series (12.4) that $f(\infty) = 0$.

Conversely, if the series (12.4) converges for $|s| > \gamma$ then (12.3) holds and the argument is reversible, to obtain the representation (12.2).

Two examples are

$$\frac{s}{s^2 - 1} = \int_0^{\infty} e^{-st} \cosh t \, dt, \qquad \sigma > 1,$$

$$(12.6) \qquad \frac{s}{s^2 + 1} = \int_0^{\infty} e^{-st} \cos t \, dt, \qquad \sigma > 0.$$

In both cases the determining function has growth $\{1, 1\}$ and the generating function is analytic for $|s| > 1$, as predicted by the theorem. We noted above that the integral (12.2) must always converge absolutely for $\sigma > \gamma$, but example (12.6) shows that the integral may at times converge in a larger half-plane. The Laplace integral (12.2) may thus sometimes serve as an analytic continuation for the series (12.4), just as series (12.4) always does so for the integral (12.2) under the conditions of the theorem.

13. The Stieltjes Transform

A transform studied by T. J. Stieltjes [1894] in connection with his investigations of infinite continued fractions is defined by

$$(13.1) \qquad f(s) = \int_0^\infty \frac{\phi(t)}{s+t}\,dt.$$

It is of special interest here because it arises naturally from the iteration of the Laplace transform:

$$\int_0^\infty e^{-sy}\,dy \int_0^\infty e^{-ty}\phi(t)\,dt = \int_0^\infty \phi(t)\,dt \int_0^\infty e^{-(s+t)y}\,dy = \int_0^\infty \frac{\phi(t)}{s+t}\,dt.$$

Under suitable conditions on $\phi(t)$ the interchange in the order of integration may be justified, but we prefer to study the transform (13.1) directly. As always, we assume that $\phi(t) \in L$ on $(0, R)$ for every $R > 0$ without further statement. We prove the following result.

Theorem 13.

$$1. \quad \int_0^\infty \frac{\phi(t)}{s_0+t}\,dt, \qquad \text{converges,} \quad \text{some } s_0 \neq 0;$$

$$2. \quad f(s) = \int_0^\infty \frac{\phi(t)}{s+t}\,dt$$

$\Rightarrow \quad f(s)$ is well defined and is analytic in the complex s-plane cut along the negative real axis.

Let D be an arbitrary compact region having no point in common with the negative real axis, $-\infty < \sigma \leq 0$, $\tau = 0$. Denote by δ the minimum distance between that segment and D and by Δ the maximum value for $|s|$, $s \in D$. Set

$$\alpha(y) = \int_0^y \frac{\phi(t)}{s_0+t}\,dt.$$

Then $\alpha(\infty) = f(s_0)$, and we have by integration by parts that

$$(13.2) \qquad f(s) = f(s_0) + (s_0 - s)\int_0^\infty \frac{\alpha(t)}{(s+t)^2}\,dt$$

for any s in D. In fact we show that the integral (13.2) represents an analytic function at interior points of D. If $R > \Delta$

$$\int_R^\infty \frac{\alpha(t)}{(s+t)^2}\,dt \ll \int_R^\infty \frac{|\alpha(t)|}{(t-\Delta)^2}\,dt < \infty,$$

$$\int_0^R \frac{\alpha(t)}{(s+t)^2}\,dt \ll \int_0^R \frac{|\alpha(t)|}{\delta^2}\,dt,$$

so that the integral (13.2) is dominated for, $s \in D$, by a convergent integral independent of s. It converges uniformly in D and consequently represents an analytic function, as stated, by a familiar theorem of Weierstrass. The proof is thus complete.

As an example, consider

$$(13.3) \qquad\qquad \frac{\pi}{\sqrt{s}} = \int_0^\infty \frac{dt}{(s+t)\sqrt{t}}.$$

This is easily proved for $\sigma > 0$ by iterating the familiar Laplace transform of $1/\sqrt{t}$. But we see by Theorem 13 and by analytic continuation of $\pi/\sqrt{s}$ that (13.3) is valid throughout the cut plane of the theorem. The example also shows that some sort of cut as that used in the theorem was essential.

14. Inversion of the Stieltjes Transform

We give here the classical inversion of the integral (13.1), originally due to Stieltjes.

Theorem 14.1.

$$(14.1) \qquad 1. \quad f(s) = \int_0^\infty \frac{\phi(t)}{s+t}\,dt \qquad \text{converges for some } s \neq 0;$$

$$2. \quad \phi(t) \in C \qquad \text{at} \quad t = t_0 > 0$$

$$(14.2) \quad \Rightarrow \quad \phi(t_0) = \lim_{\varepsilon \to 0+} \frac{f(-t_0 - i\varepsilon) - f(-t_0 + i\varepsilon)}{2\pi i}.$$

By Theorem 13 the quotient (14.2) is well defined, and

$$(14.3) \qquad \frac{f(-t_0 - i\varepsilon) - f(-t_0 + i\varepsilon)}{2\pi i} = \frac{\varepsilon}{\pi} \int_0^\infty \frac{\phi(t)\, dt}{(t - t_0)^2 + \varepsilon^2}.$$

Observe first that the theorem is true if $\phi(t) \equiv 1$. For, then

$$(14.4) \qquad \int_0^\infty \frac{\varepsilon\, dt}{(t - t_0)^2 + \varepsilon^2} = \cot^{-1}\left(\frac{-t_0}{\varepsilon}\right) \to \pi, \qquad \varepsilon \to 0+.$$

Hence without loss of generality we may assume $\phi(t_0) = 0$. Now choose $R > t_0$ and express the integral (14.3) as the sum of two others $I_1(\varepsilon)$ and $I_2(\varepsilon)$ corresponding to the two intervals $(0, R)$ and (R, ∞), respectively. Then if $0 < t_0 - \delta < t_0 + \delta < R$,

$$|I_1(\varepsilon)| \leqq \sup_{|t-t_0|<\delta} |\phi(t)| \frac{\varepsilon}{\pi} \int_{|t-t_0|<\delta} \frac{dt}{(t - t_0)^2 + \varepsilon^2}$$

$$+ \frac{\varepsilon}{\pi} \int_{\substack{0 \\ |t-t_0|>\delta}}^{R} \frac{|\phi(t)|\, dt}{(t - t_0)^2}.$$

By (14.4) we have

$$\varlimsup_{\varepsilon \to 0} |I_1(\varepsilon)| \leqq \sup_{|t-t_0|>\delta} |\phi(t)|.$$

Consequently, $I_1(0+) = 0$, by hypothesis 2. The integral $I_2(\varepsilon)$ need not converge absolutely. Accordingly we introduce the function

$$\alpha(t) = \int_0^t \phi(y)\, dy.$$

Integration by parts gives

$$I_2(\varepsilon) = \frac{-\varepsilon\alpha(R)}{\pi[(R - t_0)^2 + \varepsilon^2]} + \frac{2\varepsilon}{\pi} \int_R^\infty \frac{\alpha(t)(t - t_0)}{[(t - t_0)^2 + \varepsilon^2]^2}\, dt,$$

$$(14.5) \qquad |I_2(\varepsilon)| \leqq \frac{\varepsilon|\alpha(R)|}{\pi(R - t_0)^2} + \frac{2\varepsilon}{\pi} \int_R^\infty \frac{|\alpha(t)|}{(t - t_0)^3}\, dt,$$

and $I_2(0+) = 0$ also, provided only that the integral on the right of (14.5) converges. By Theorem 14 the integral (14.1) converges at $s = 1$, for example, so that the function

$$\beta(t) = \int_0^t \frac{\phi(y)\, dy}{1 + y}$$

is bounded on $(0, \infty)$. Integration by parts gives

$$\alpha(t) = \beta(t)(t + 1) - \int_0^t \beta(t)\,dt = O(t), \qquad t \to \infty.$$

Hence the convergence of the integral on the right of (14.5) is immediate, and the proof is complete.

In formula (14.2) the variable s of the function $f(s)$ approaches $-t_0$ from above and below along the vertical line $\sigma = -t_0$. A more general approach is possible, as we now show. Let $o(1)$ be a real function of ε which approaches zero with ε. Then it may be shown that

$$(14.6) \qquad \lim_{\varepsilon \to 0+} \frac{f(-t_0 + o(1) - i\varepsilon) - f(-t_0 + o(1) + i\varepsilon)}{2\pi i} = \phi(t_0).$$

The previous proof goes through, *mutatis mutandis*. We omit details but observe that (14.4) becomes

$$\int_0^\infty \frac{\varepsilon\,dt}{(t - t_0 + o(1))^2 + \varepsilon^2} = \cot^{-1}\left(\frac{-t_0 + o(1)}{\varepsilon}\right) \to \pi, \qquad \varepsilon \to 0+.$$

Accordingly we must now prove that

$$\varepsilon \int_0^\infty \frac{\phi(t)\,dt}{(t - t_0 + o(1))^2 + \varepsilon^2} \to 0, \qquad \varepsilon \to 0+,$$

and this is accomplished much as before.

By use of (14.6) we can now prove that the approach of s to $-t_0$ may be along a circular arc, a result we shall need in Chapter 9.

Corollary 14.1.

$$(14.7) \qquad \lim_{\eta \to 0+} \frac{f(t_0\,e^{-i(\pi - \eta)}) - f(t_0\,e^{i(\pi - \eta)})}{2\pi i} = f(t_0).$$

For, if we relate the variables ε and η of theorem and corollary, respectively, by the equation

$$\varepsilon = t_0 \sin \eta,$$

then they tend to zero together and the function $o(1)$ of equation (14.6) becomes

$$o(1) = t_0(1 - \cos \eta).$$

This function does indeed tend to zero with ε and η, and (14.6) becomes (14.7), as desired.

As an example let us use Corollary 14.1 to invert the transform (13.3). Equation (14.7) becomes

$$\frac{t_0^{-1/2}[e^{i(\pi-\eta)/2} - e^{-i(\pi-\eta)/2}]}{2i} = t_0^{-1/2}\cos\frac{\pi\eta}{2} \to \frac{1}{(t_0)^{1/2}} \qquad \eta \to 0+,$$

and we have recovered the determining function.

In Chapter 9 we shall be considering in some detail the potential transform

$$f(s) = \frac{2}{\pi}\int_0^\infty \frac{t\phi(t)}{s^2 + t^2}\,dt.$$

Since it is equivalent to the Stieltjes transform

$$f(\sqrt{s}) = \frac{1}{\pi}\int_0^\infty \frac{\phi(\sqrt{t})}{s+t}\,dt,$$

we can already obtain an inversion of it by use of Corollary 14.1.

Theorem 14.2.

1. $\displaystyle f(s) = \frac{2}{\pi}\int_0^\infty \frac{t\phi(t)}{s^2 + t^2}\,dt$ converges for some $s \neq 0$;

2. $\phi(t) \in C$ at $t = t_0 > 0$

$$(14.8) \quad \Rightarrow \quad \lim_{\varepsilon \to 0+} \frac{f(t_0 e^{-i(\pi/2-\varepsilon)}) - f(t_0 e^{i(\pi/2-\varepsilon)})}{2i} = \phi(t_0).$$

As an example, consider the pair $\phi = 1/t$, $f = 1/s$. In this case equation (14.8) becomes

$$\lim_{\varepsilon \to 0+} \frac{\cos\varepsilon}{t_0} = \frac{1}{t_0},$$

and the inversion is verified.

15. Summary

The summary at the end of Chapter 2 would serve equally well here. The main additional material involves the matter of representation.

Let us list the basic inversion formulas and the Hamburger conditions for representation. If

$$f(s) = \int_0^\infty e^{-st}\, d\alpha(t),$$

then

(15.1)
$$\alpha(t) = \frac{1}{2\pi i} \int_{c-i\infty}^{c+i\infty} \frac{f(s)e^{st}}{s}\, ds.$$

If

$$f(s) = \int_0^\infty e^{-st}\phi(t)\, dt,$$

then

(15.2)
$$\phi(t) = \frac{1}{2\pi i} \int_{c-i\infty}^{c+i\infty} f(s)e^{st}\, ds.$$

The integrals (15.1) and (15.2) may diverge, but the Cauchy principle value is always effective. Note that (15.2) follows formally from (15.1) by differentiation under the sign, as one would expect since $\phi(t)$ is equal to the derivative of $\alpha(t)$ almost everywhere.

The Hamburger sufficient conditions for the validity of

$$f(s) = \int_{-\infty}^{\infty} e^{-st}\phi(t)\, dt$$

with the integral converging absolutely in the strip $\alpha < \sigma < \beta$ are that (1) $f(s) \in A$ there, (2) $f(s) \in L$ on each vertical line of the strip, and (3) that $f(\sigma + iy) = o(1)$ uniformly in $\alpha < \sigma < \beta$ as $|y| \to \infty$.

If β is replaced by $+\infty$ in conditions (1) and (2), and if (3) is replaced by (3′) $f(s) = o(1)$ as $|s| \to \infty$ in the half-plane $\sigma > \alpha$, the conditions become sufficient for the validity of

$$f(s) = \int_0^\infty e^{-st}\phi(t)\, dt, \qquad \sigma > \alpha.$$

Finally it is useful to recall that any function analytic at infinity and vanishing there is always a Laplace transform of an entire function of predictable maximum growth.

EXERCISES

1. Let z and s be complex constants with positive real parts, w a complex variable. Prove by Cauchy's theorem that the integral of $e^{-w}w^z$ has the same value over the two paths
    (a) $w = st,$ $0 \leq t < \infty,$
    (b) $w = t,$ $0 \leq t < \infty.$
    Thus prove that

    $$(16.1) \qquad \frac{\Gamma(z)}{s^z} = \int_0^\infty e^{-st}t^{z-1}\, dt, \qquad \text{Re } z > 0, \quad \text{Re } s > 0.$$

2. Show that the integral (16.1) converges at $s = i$ if $0 < \text{Re } z < 1$. Obtain its value from (16.1) and Theorem 4.1. Thus establish formulas E and F of §2, at least for $0 < \sigma < 1$. Complete the proof of formula E by analytic continuation, using Theorem 4.2. In particular check E at $s = 0$, $s = -1/2$.

3. Prove formula B of §2 by expressing $1/(x^2 + 1)$ as a Laplace integral and reversing the order of integration in the resulting iterated integral. Justify by Fubini's theorem, at least for $1 < \sigma < 2$. Complete the proof by analytic continuation.

4. Prove formula C of §2 by change of variable in B.

5. Prove formula G of §2 by completing the square in the exponents of the integrand.

6. Use Corollary 8 to express $\Gamma(s/2)\Gamma(1 - (s/2))$ as a Mellin transform and thus get a new proof of formula B, §2.

7. Show that the capital O of Lemma 5.2 may be replaced by small o.

8. Find a Mellin transform representation for $\Gamma(s)$ valid in $-1 < \sigma < 0$; in $-2 < \sigma < -1$; ... in $-n - 1 < \sigma < -n$. [Hint: $\Gamma(s) = \Gamma(s + 1)/s$ and

    $$\frac{1}{s} = -\int_1^\infty x^{s-1}\, dx, \qquad \sigma < 0.]$$

9. If x, δ, ρ are positive constants, find

    $$\underset{0 \leq t \leq x}{\text{Max}} \quad [(x + \delta - t)^\rho - (x - t)^\rho],$$

    thus verifying inequalities (9.3).

10.   If $\alpha < 0$, $\beta < 0$, show $D^{\beta}(D^{\alpha}f(x)) = D^{\alpha+\beta}f(x)$. What are you assuming about $f(x)$? Discuss $D^{\alpha}D^{-\alpha}$ and $D^{-\alpha}D^{\alpha}$.

11.   From equation (2.1) of Chapter 3 show that

$$\zeta(s) = s \int_0^1 x^{s-1} [1/x] \, dx, \qquad \sigma > 1.$$

By Corollary 8 express $\Gamma(s)\zeta(s)$ as a Mellin transform thus producing a new proof of Lemma 7.2 of Chapter 3. $\omega(x)$ will be

$$\omega(x) = x \int_1^{\infty} e^{-xy}[y] \, dy = \frac{1}{e^x - 1}.$$

12.   Show that $f(s) = 1/(1 - s^2)$ satisfies the hypotheses of Theorem 11.1 in $-1 < \sigma < 1$ and find the corresponding determining function.

$$\text{Ans.} \quad \phi(t) = \frac{e^{-|t|}}{2}.$$

13.   Show that any function analytic at ∞ is a Laplace–Stieltjes transform.

14.   Show that $f(s) = 1/(1 - s^2)$ satisfies the conditions of Theorem 12 with $\gamma = 1$ and find the corresponding determining function.

$$\text{Ans.} \quad \phi(t) = -\sinh t.$$

15. Find by two methods the determining function corresponding to the generating function $1/[s(s^2 + 1)]$.

16.   Find $\phi(x)$ so that

$$\frac{s}{\sin \pi s/2} = \int_0^{\infty} x^{s-1}\phi(x) \, dx, \qquad |\sigma| < 2.$$

$$\text{Ans.} \quad \frac{4}{\pi} \frac{x^2}{(x^2 + 1)^2}.$$

6 *Real Inversion Theory*

1. Introduction

In §7 of Chapter 5 we gave the classical inversion of the Laplace transform and in §14 we gave the analogous inversion of the Stieltjes transform. Both involve the determining functions for complex values of the variable. There are two more recent inversions, studied extensively by the author, [1934] and [1938], which involve only real values of the variable. In Chapter 7 we shall develop these real inversions as special cases of a much more general theory, but in the present chapter we treat them more expeditiously, using the specific properties of the kernels involved. The chief analytic equipment needed for this direct approach is the Laplace asymptotic method, and we shall begin with a rapid development thereof. It involves the estimation of integrals of the form

$$\int_a^b [g(t)]^n \phi(t)\, dt$$

for large values of the parameter n. It is important that the function $g(t)$ should be positive, for the positiveness enables one to alleviate the conditions to be imposed on $\phi(t)$. In our application to the inversion of the Laplace transform we are thus able to relax the restrictive local hypothesis on the determining function, such as hypothesis 2 of Theorem 7.3, Chapter 5.

2. Laplace's Asymptotic Method

A very useful tool, indispensable for the work of this chapter, is Laplace's method of estimating the behavior of certain integrals for large values of the parameter. In most cases these integrals cannot be explicitly evaluated. We shall begin by giving a heuristic approach, illustrating and checking the method by use of integrals which *can* be evaluated.

Consider an integral in which the positive parameter n (which need not be an integer) occurs as a power of one factor of the integrand:

$$I_n = \int_a^b [g(t)]^n \phi(t)\, dt = \int_a^b e^{nh(t)} \phi(t)\, dt.$$

It is required to obtain the asymptotic behavior of I_n as $n \to \infty$. That is, we seek a function A_n such that

$$\lim_{n \to \infty} \frac{I_n}{A_n} = 1.$$

Since large values of a power get larger as the power increases and small values get smaller it would be natural to expect that the maximum value of g or of h in (a, b) would play an important rôle in the solution. This is in fact the case. If there are several points of (a, b) where the maximum is taken on we can break the integral into several parts, in each of which the maximum is at an end of the interval of integration. Let us consider a variety of such cases.

In the present section let us assume that $h(t)$ and $\phi(t)$ have sufficient properties for the Taylor expansions used. Later we shall make all conditions precise.

Case I

$$h(t) \in\, \downarrow \text{ in } (a, b); \quad h'(a) < 0; \quad \phi(a) \neq 0.$$

Then

$$(2.1) \qquad\qquad I_n \sim \frac{-\phi(a)\, e^{nh(a)}}{[nh'(a)]}, \qquad n \to \infty.$$

To see why this is a reasonable result, note first that the first factor in the integrand of I_n is largest at $t = a$. Consequently we replace each factor of the integrand by the leading term in its Taylor development:

$$I_n \sim \int_a^b e^{n[h(a)+h'(a)(t-a)]}\phi(a)\, dt = \phi(a)e^{nh(a)}\int_0^{b-a} e^{nh'(a)t}\, dt.$$

This integral can be evaluated, but its size is not greatly changed if the upper limit of integration is replaced by ∞. It then becomes a Laplace integral whose value is $1/[-nh'(a)]$. We could now establish (2.1) rigorously following the method just sketched. We shall not do so in this simple case, since there will be no application of it made later. Let us rather illustrate it by the example

$$(2.2) \quad I_n = \int_0^1 (1 - t)^n(2 + 3t)\, dt = \frac{2n + 7}{(n + 1)(n + 2)} \sim \frac{2}{n}, \qquad n \to \infty.$$

Using (2.1) we have $a = 0$, $h(0) = 0$, $h'(0) = -1$, $\phi(0) = 2$, so that (2.1) and (2.2) agree.

Case II

$$h(t) \in \downarrow \text{ in } (a, b); \quad h'(a) < 0; \quad \phi(a) = 0; \quad \phi'(a) \neq 0.$$

Then

$$(2.3) \qquad\qquad I_n \sim \frac{\phi'(a)\, e^{nh(a)}}{[nh'(a)]^2}, \qquad n \to \infty.$$

Proceeding as in Case I, we are led to the estimate

$$\phi'(a)e^{nh(a)}\int_0^\infty e^{nh'(a)t}t\, dt = \frac{\phi'(a)e^{nh(a)}}{[-nh'(a)]^2}.$$

As an example take

$$(2.4) \qquad I_n = \int_0^2 (2 - t)^n t\, dt = \frac{2^{n+2}}{(n + 1)(n + 2)} \sim \frac{2^{n+2}}{n^2}, \qquad n \to \infty.$$

Here $h(a) = h(0) = \log 2$, $h'(0) = -\frac{1}{2}$, $\phi(0) = 0$, $\phi'(0) = 1$, so that (2.3) and (2.4) agree.

Case III

$$h(t) \in \downarrow \text{ in } (a, b); \quad h'(a) = 0; \quad h''(a) < 0; \quad \phi(a) \neq 0.$$

Then

(2.5) $$I_n \sim \sqrt{\pi} \, \frac{e^{nh(a)}\phi(a)}{[-2nh''(a)]^{1/2}}, \qquad n \to \infty.$$

This is the case which we shall find most useful in subsequent work. This time we need one more term in Taylor's expansion of $h(t)$ and thus arrive at the estimate

$$\phi(a)e^{nh(a)} \int_0^\infty \exp\left(\frac{nh''(a)t^2}{2}\right) dt,$$

and this is equal to the right-hand side of (2.4) by use of the probability integral

$$\int_0^\infty \exp(-t^2) \, dt = \frac{1}{2}\sqrt{\pi}.$$

For our illustration here we choose

$$I_n = \int_0^1 (1 - t^2)^n \, dt = \frac{1}{2} B\left(n + 1, \frac{1}{2}\right) = \frac{4^n n! n!}{(2n + 1)(2n)!}.$$

By use of Stirling's formula

$$I_n \sim \left(\frac{\pi}{4n}\right)^{1/2}, \qquad n \to \infty.$$

To apply formula (2.5) we have $h(t) = \log(1 - t^2)$, $\phi(t) = 1$, $h(0) = 0$, $h'(0) = 0$, $h''(0) = -2$. Hence it gives the same result.

We can now impose a set of sufficient conditions for the validity of the result conjectured in Case III above.

Theorem 2.1.

1. $\phi(t) \in L, \qquad a \leq t \leq b, \quad \phi(a+)$ exists;
2. $h(t) \in C^2 . \downarrow, \qquad a \leq t \leq b, \quad h'(a) = 0, \quad h''(a) < 0$

(2.6) $\Rightarrow$ $$I_n = \int_a^b e^{nh(t)}\phi(t) \, dt \sim \sqrt{\pi}\,\phi(a+)e^{nh(a)}[-2nh''(a)]^{-1/2},$$

$$n \to \infty.$$

Since we have not postulated here that $\phi(a+) \neq 0$, (2.6) must be understood to mean that

$$I_n = o\left(\frac{e^{nh(a)}}{\sqrt{n}}\right), \qquad n \to \infty$$

when $\phi(a+) = 0$.

By use of a linear transformation we may assume that $a = 0$. Again, by multiplying (2.6) by $e^{nh(a)}$ we see that we may take $h(a) = 0$ without loss of generality. Let us first treat the special case $\phi(t) \equiv 1$. We must show

$$\lim_{n \to \infty} \sqrt{n}\, I_n = \pi^{1/2}\, [-2h''(0)]^{-1/2}.$$

For an arbitrary positive $\varepsilon < -h''(0)$ there corresponds a δ such that when $0 \leq t \leq \delta$

$$-2q^2 = h''(0) - \varepsilon \leq h''(t) \leq h''(0) + \varepsilon = -2p^2.$$

Here we have introduced positive numbers p and q as abbreviations. They both approach $[-h''(0)/2]^{1/2}$ as $\varepsilon \to 0$. By Taylor's formula

$$h(t) = h''(\xi) \frac{t^2}{2}, \qquad 0 < \xi < t,$$

so that

(2.7) $$-q^2 t^2 \leq h(t) \leq -p^2 t^2, \qquad 0 \leq t \leq \delta.$$

Since $h(t) \in \downarrow$

$$\sqrt{n} \int_\delta^b e^{nh(t)}\, dt \leq \sqrt{n}\, e^{nh(\delta)}(b - \delta) = o(1), \qquad n \to \infty.$$

Here we have used the fact that $h(\delta) < 0$ by virtue of (2.7). Hence

$$\sqrt{n} \int_0^\delta \exp(-nq^2 t^2)\, dt + o(1) \leq \sqrt{n}\, I_n \leq \sqrt{n} \int_0^\delta \exp(-np^2 t^2)\, dt + o(1)$$

$$\frac{1}{q} \int_0^{\sqrt{n}q\delta} \exp(-t^2)\, dt + o(1) \leq \sqrt{n}\, I_n \leq \frac{1}{p} \int_0^{\sqrt{n}p\delta} \exp(-t^2)\, dt + o(1).$$

Allowing n to become infinite we see, by use of the probability integral, that

$$\frac{\sqrt{\pi}}{2q} \leqq \varliminf_{n \to \infty} \sqrt{n}\, I_n \leqq \varlimsup_{n \to \infty} \sqrt{n}\, I_n \leqq \frac{\sqrt{\pi}}{2p}.$$

The two inner terms of this continued inequality do not depend on ε, while the two outer ones approach the same value as $\varepsilon \to 0$. Hence

$$(2.8) \qquad \lim_{n \to \infty} \sqrt{n}\, I_n = \sqrt{\pi}\,(-2h''(0))^{-1/2},$$

as desired.

Now in the general case, when $\phi(t)$ is not constant, it will be enough to show that

$$\lim_{n \to \infty} \sqrt{n} \int_0^b e^{nh(t)}[\phi(t) - \phi(0+)]\, dt = 0,$$

or to assume that $\phi(0+) = 0$ and to show that

$$\lim_{n \to \infty} \sqrt{n} \int_0^b e^{nh(t)}\phi(t)\, dt = 0.$$

But for any δ between 0 and b

$$\sqrt{n}\,|I_n| \leqq \sup_{0 \leqq t \leqq \delta} |\phi(t)| \sqrt{n} \int_0^\delta e^{nh(t)}\, dt + \sqrt{n}\, e^{nh(\delta)} \int_\delta^b |\phi(t)|\, dt.$$

Hence

$$\varlimsup_{n \to \infty} \sqrt{n}\,|I_n| \leqq \sup_{0 \leqq t \leqq \delta} |\phi(t)|\, c,$$

where c is the constant (2.8). Now as δ approaches zero the right side does also, and the proof is complete.

We now make several immediate extensions of the previous theorem. If the maximum value of $h(t)$ occurs at b rather than at a, then a simple change of variable shows that

$$(2.9) \qquad \int_a^b e^{nh(t)}\phi(t)\, dt \sim \sqrt{\pi}\,\phi(b-)e^{nh(b)}[-2nh''(b)]^{-1/2}, \qquad n \to \infty.$$

Or, if the maximum is at an intermediate point c, then

$$(2.10) \qquad \int_a^b e^{nh(t)}\phi(t)\, dt \sim 2\sqrt{\pi}\,\phi(c)e^{nh(c)}[-2nh''(c)]^{-1/2}, \qquad n \to \infty.$$

In many applications of the Laplace method the integral I_n is extended over an infinite range. To treat such cases we state and prove the following theorem.

Theorem 2.2.

1.   Hypotheses 1 and 2 of Theorem 2.1 hold for every $b > a$;

2. $\displaystyle\int_a^\infty e^{mh(t)}|\phi(t)|\,dt < \infty,$ some m

$$(2.11) \quad \Rightarrow \int_a^\infty e^{nh(t)}\phi(t)\,dt \sim \sqrt{\pi}\,\phi(a+)\,e^{nh(a)}[-2nh''(a)]^{-1/2}, \quad n \to \infty.$$

Again reducing to the case $a = 0$, $\phi(0+) = 0$, we must show that

$$\lim_{n\to\infty} \sqrt{n}\,I_n = \lim_{n\to\infty} \sqrt{n}\int_0^\infty e^{nh(t)}\phi(t)\,dt = 0.$$

But for $n > m$ and $b > a$

$$\left|\sqrt{n}\int_b^\infty e^{nh(t)}\phi(t)\,dt\right| \leqq \sqrt{n}\,e^{(n-m)h(b)}\int_b^\infty e^{mh(t)}|\phi(t)|dt.$$

Since $h(b) < 0$ the right side $\to 0$ as $n \to \infty$. Hence

$$\sqrt{n}\,I_n = \sqrt{n}\int_0^b e^{nh(t)}\phi(t)\,dt + o(1), \quad n \to \infty.$$

But the first term on the right also $\to 0$ as $n \to \infty$ by Theorem 2.1. This completes the proof.

As an application let us prove Stirling's formula:

$$(2.12) \qquad n! \sim n^n e^{-n}\sqrt{2\pi n}, \quad n \to \infty.$$

By a familiar Laplace transform

$$n! = n^{n+1}\int_0^\infty (e^{-t}t)^n\,dt.$$

Apply (2.9) for the interval $(0, 1)$ and (2.11) for interval $(1, \infty)$. In both cases $h(t) = -t + \log t$, $\phi(t) = 1$. We may choose $m = 1$ since

$$\int_1^\infty e^{-t}t\,dt < \infty.$$

Hence $h(1) = -1$, $h'(1) = 0$, $h''(1) = -1$, and (2.10) becomes

$$(2.13) \qquad 2\sqrt{\pi}\, e^{nh(1)}[-2nh''(1)]^{-1/2} = 2\sqrt{\pi}\, \frac{e^{-n}}{\sqrt{2n}},$$

so that (2.12) is proved.

3. Real Inversion of the Laplace Transform

There are two classical inversions of a power series, one the Cauchy method of contour integration, the other the Taylor method using derivatives. We may expect analogous methods for the Laplace transform. In the previous chapter we have already found for the Laplace integral an inversion by contour integration. We now obtain one by use of derivatives. In the Taylor method for inverting the series,

$$(3.1) \qquad F(x) = \sum_{k=0}^{\infty} a_k x^k, \qquad a_k = \frac{F^{(k)}(0)}{k!},$$

all derivatives of F at $x = 0$ are used. These could all be computed from the values of $F(x)$ in a neighborhood $(0, \delta)$ of the origin. To convert (3.1) to a Dirichlet series, the prototype for a Laplace transform, we replace x by e^{-x} thus converting the neighborhood $(0, \delta)$ into $(\log 1/\delta, \infty)$. Thus we would look for an operator which would utilize the values of the generating function in a neighborhood of $+\infty$ and which presumably would involve all of its derivatives. We now define such an operator.

Definition 3. For integers $k > 0$

$$L_{k,y}[f(x)] = \frac{(-1)^k}{k!} \left(\frac{k}{y}\right)^{k+1} f^{(k)}\left(\frac{k}{y}\right).$$

Note that this operator will be defined for *any* positive y and *all* large k if $f(x) \in C^\infty$ in a neighborhood (however small) of $+\infty$. We shall see that as $k \to \infty$ it does generally invert the Laplace–Lebesgue integral.

For example,

$$(3.2) \qquad L_{k,y}\left[\frac{1}{x^2}\right] = \frac{k}{k+1}y,$$

$$(3.3) \qquad L_{k,y}\left[\frac{1}{\sqrt{x}}\right] = \binom{2k}{k}\frac{1}{4^k}\sqrt{\frac{k}{y}}.$$

If $f(x)$ has an abscissa of absolute convergence σ_a, we shall show that the operator of Definition 3 does indeed approach the determining function as $k \to \infty$ at all points of continuity thereof.

Theorem 3.

$$(3.4) \qquad 1. \quad f(x) = \int_0^\infty e^{-xt}\phi(t)\, dt, \qquad x > \sigma_a,$$

$$2. \quad \phi(t) \in C \text{ at } t = y > 0$$

$$(3.5) \quad \Rightarrow \qquad \lim_{k \to \infty} L_{k,y}[f(x)] = \phi(y).$$

We know from Chapter 5 that differentiation under the integral sign is valid in the region of convergence. Hence

$$L_{k,y}[f(x)] = \frac{1}{k!}\left(\frac{k}{y}\right)^{k+1}\int_0^\infty e^{-kt/y}t^k\phi(t)\, dt$$

when $k/y > \sigma_a$. After change of variable this equation becomes

$$(3.6) \qquad L_{k,y}[f(x)] = \frac{k^{k+1}}{k!}\int_0^\infty (e^{-t}t)^k\phi(ty)\, dt,$$

and we are in a position to apply Laplace's asymptotic method. We may choose the constant m of Theorem 2.2 as any number greater than $y\sigma_a$; for, then

$$\int_1^\infty (e^{-t}t)^m\,|\phi(ty)|\, dt < \infty.$$

But now the computations are the same as those used in §2 for the proof of Stirling's formula. Consequently the integral (3.6) (without the exterior factor) is asymptotic to $\phi(y)\sqrt{2\pi}\,e^{-k}/\sqrt{k}$. This is equation (2.13). Now taking into account the factor $k^{k+1}/k!$, we have the desired result.

As examples, consider

$$\frac{1}{x^2} = \int_0^\infty e^{-xt} t \, dt, \qquad \sigma_a = 0,$$

$$\frac{1}{\sqrt{x}} = \int_0^\infty \frac{e^{-xt}}{\sqrt{\pi t}} \, dt, \qquad \sigma_a = 0.$$

It is easy to compute the limits of the functions (3.2) and (3.3) to obtain y and $1/(\pi y)^{1/2}$, respectively, for all $y > 0$, as predicted by the theorem.

The results of the previous section can be greatly extended. We have postulated the absolute convergence of the given transform and the continuity of the determining function in order to keep proofs simple. However, it is known that if (3.4) *converges somewhere* and if $\phi(t) \in L$ in $(0, R)$ for every $R > 0$, then (3.5) holds almost everywhere. The present inversion should be contrasted with Theorem 7.3 of Chapter 5 where the strong local condition $\phi(t) \in V$ was required. That was needed on account of the intervention of the second mean-value theorem, which can be avoided here since the kernel $(e^{-t}t)^k$ is positive.

4. The Stieltjes Transform

As a second application of the Laplace method let us seek an inversion of the *Stieltjes transform*,

$$(4.1) \qquad f(x) = \int_0^\infty \frac{\phi(t)}{x + t} \, dt = \lim_{R \to \infty} \int_0^R \frac{\phi(t)}{x + t} \, dt.$$

We shall assume that $\phi(t) \in L$ in $0 \leq t \leq R$ for every R so that the transform is defined when the limit (4.1) exists. We note first that if it exists for one positive value $x = x_0$ it does so for all positive values. See Theorem 13 of Chapter 5. Similarly, if (4.1) converges absolutely at $x_0 > 0$, it does so for all $x > 0$.

If (4.1) converges absolutely for $x > 0$, then by Fubini's theorem it is a Laplace transform of the Laplace transform of ϕ,

$$(4.2) \qquad f(x) = \int_0^\infty e^{-xt} \, dt \int_0^\infty e^{-ty} \phi(y) \, dy.$$

We thus see from Chapter 5 that (4.1) can be differentiated as often as desired under the integral sign.

We now introduce a Linear differential operator, somewhat analogous to that of Definition 3.

Definition 4. For integers $k > 0$

$$M_{k,y}[f(x)] = \frac{(-1)^{k+1}}{k!\,k!}\, D^k[y^{2k+1}D^{k+1}f(y)], \qquad D = \frac{d}{dy}.$$

We shall show that this operator plays the same rôle for the Stieltjes transform as that of Definition 3 did for the Laplace. For example,

$$M_{2,y}[f] = -\frac{1}{4}\,[y^5 f^{(5)} + 10y^4 f^{(4)} + 20y^3 f^{(3)}].$$

This is seen to be an Euler differential operator; that is, a linear one in which each coefficient is a power of y of degree equal to the order of the corresponding derivative. This is true for all k. For arbitrary k it could be characterized as the linear differential operator of order $2k + 1$ with fundamental solutions

$$x^n, n = 0, \pm 1, \pm 2, \ldots, \pm k$$

and with leading coefficient

$$(-1)^{k+1} y^{2k+1}/(k!)^2.$$

As an example,

$$(4.3) \qquad M_{k,y}\left[\frac{1}{\sqrt{x}}\right] = \frac{2(2k+1)!(2k)!}{4^{2k+1}(k!)^4}\,\frac{1}{\sqrt{y}}.$$

As another example let us apply the operator to the function $(x + t)^{-1}$, the kernel of the Stieltjes transform. To facilitate the computation we use a simple lemma about homogeneous functions of two variables.

Lemma 4.1.

$$1. \quad f(x, t) \in C^2, \qquad x, t > 0;$$

$$2. \quad f(\lambda x, \lambda t) = \lambda^{-1} f(\lambda x, \lambda t), \qquad \text{all } \lambda > 0$$

$$\Rightarrow \qquad \frac{\partial}{\partial x} f(x, t) = -\frac{\partial}{\partial t}\frac{t}{x} f(x, t), \qquad x, t > 0.$$

For, $tf(x, t)$ is homogeneous of degree zero. By Euler's theorem

$$x \frac{\partial}{\partial x}[tf(x, t)] + t \frac{\partial}{\partial t}[tf(x, t)] = 0,$$

and this is equivalent to the stated result.

Lemma 4.2. For $x > 0, t > 0$

$$M_{k,y}\left[\frac{1}{x+t}\right] = \frac{(2k+1)!}{k!k!} \frac{t^k y^{k+1}}{(y+t)^{2k+2}}.$$

For,

$$(-1)^{k+1} \frac{\partial^{k+1}}{\partial y^{k+1}} \frac{1}{y+t} = \frac{(k+1)!}{(y+t)^{k+2}}.$$

Multiply by y^{2k+1}/t^k to obtain a homogeneous function of order -1 and apply Lemma 4.1 k times. At each stage the factor t/y must be inserted, so that

$$M_{k,y}\left[\frac{1}{x+t}\right] = \frac{(-t)^k(k+1)!}{k!\,k!} \frac{\partial^k}{\partial t^k} \frac{y^{k+1}}{(y+t)^{k+2}}.$$

Performing the indicated differentiation now yields the stated result.

We can now invert the Stieltjes transform.

Theorem 4.

1. $f(x) = \int_0^\infty \frac{\phi(t)}{x+t}\, dt,$ converges absolutely $x > 0$;

2. $\phi(t) \in C$ at $t = y > 0$

$\Rightarrow$ $\displaystyle\lim_{k\to\infty} M_{k,y}[f(x)] = \phi(y).$

As we have noted above, differentiation under the integral sign is valid, so that Lemma 4.2 yields

$$M_{k,y}[f(x)] = \frac{(2k+1)!}{k!k!} \int_0^\infty \frac{t^k y^{k+1}}{(y+t)^{2k+2}}\, \phi(t)\, dt$$

$$(4.4) \qquad = \frac{(2k+1)!}{k!k!} \int_0^\infty \left[\frac{t}{(t+1)^2}\right]^k \frac{\phi(ty)}{(t+1)^2}\, dt.$$

We now apply Laplace's asymptotic method, using (2.9) for the interval $(0, 1)$ and (2.11) for $(1, \infty)$. In both cases $h(t) = \log t - 2 \log (t + 1)$, $h(1) = - \log 4$, $h'(1) = 0$, $h''(1) = - \frac{1}{2}$. Thus the integral (4.4) (without the external numerical factor) is asymptotic to

$$\frac{2\sqrt{\pi}\,\phi(y)}{\sqrt{k}\,4^{k+1}}, \qquad k \to \infty.$$

Now using Stirling's formula we have

$$\frac{(2k + 1)!}{k!\,k!} \sim 2^{2k+1}\sqrt{\frac{k}{\pi}}, \qquad k \to \infty,$$

so that the theorem is proved. Note that the constant m of Theorem 2.2 may here be chosen as zero since

$$\int_1^\infty \frac{|\phi(ty)|}{(t + 1)^2}\,dt \leqq \frac{1}{2}\int_y^\infty \frac{|\phi(t)|}{t + y}\,dt.$$

As an example take

$$\frac{1}{\sqrt{x}} = \frac{1}{\pi}\int_0^\infty \frac{dt}{\sqrt{t(x + t)}}.$$

This may be verified by equation (4.2) with $\phi(y) = y^{-1/2}$. But now it is easy to compute directly the limit of the right-hand side of (4.3) as $k \to \infty$ to obtain

$$\lim_{k \to \infty} M_{k,\,y}\left[\frac{1}{\sqrt{x}}\right] = \frac{1}{\pi\sqrt{y}},$$

as predicted by the theorem.

5. The Hausdorff Moment Problem; Uniqueness

In his study of methods for summing divergent series F. Hausdorff [1921] was led to inquire what sequences μ_n can be written as

$$(5.1) \qquad \mu_n = \int_0^1 t^n\,d\alpha(t), \qquad n = 0, 1, 2, \ldots$$

with $\alpha(t) \in \uparrow$. The sequence $\mu_n = 1/(n+1)$ has the property $(\alpha(t) = t)$, whereas any unbounded sequence does not. Hausdorff found a neat characterization of those that do. We present it here because of its usefulness in obtaining a representation theorem for the Laplace integral.

We show first that a sequence cannot have more than one representation (5.1) by a normalized function $\alpha(t)$ of bounded variation. Recall the definition of V^* from §7 of Chapter 5:

$$(5.2) \qquad \alpha(0) = 0, \qquad \alpha(t) = \frac{\alpha(t+) + \alpha(t-)}{2}, \qquad 0 < t < 1.$$

The uniqueness will follow from the next theorem.

Theorem 5.1.

1. $\alpha(t) \in V^*, \qquad 0 \leqq t \leqq 1;$

2. $\displaystyle\int_0^1 t^n \, d\alpha(t) = 0, \qquad n = 0, 1, 2, \ldots$

$$\Rightarrow \qquad\qquad \alpha(t) = 0, \qquad 0 \leqq t \leqq 1.$$

Two integrations by parts show that hypothesis 2 implies

$$(5.3) \qquad\qquad \int_0^1 t^n \beta(t) \, dt = 0, \qquad n = 0, 1, 2, \ldots$$

where

$$(5.4) \qquad\qquad \beta(t) = \int_0^t \alpha(y) \, dy.$$

By (5.4) $\beta(t)$ is continuous together with its conjugate $\bar\beta(t)$. (We are here assuming, for increased generality, that $\alpha(t)$ is complex.) By Weierstrass's theorem of approximation there corresponds to every $\varepsilon > 0$ a polynomial $P(t)$ such that $|\bar\beta(t) - P(t)| < \varepsilon$ for $0 \leqq t \leqq 1$. By equations (5.3)

$$(5.5) \qquad \int_0^1 |\beta(t)|^2 \, dt = \int_0^1 \beta(t)[\bar\beta(t) - P(t)] \, dt \leqq \varepsilon \int_0^1 |\beta(t)| \, dt.$$

The left side, being independent of ε, can only be zero if (5.5) is to hold for all ε. That is, $\beta(t)$ and $\beta'(t)$ are identically zero. But $\beta'(t) = \alpha(t)$ at the dense set of points of continuity of $\alpha(t)$. Then for $0 < t < 1$, $\alpha(t+) = \alpha(t-) = 0$, because these one-sided limits are known to exist for $\alpha(t) \in V$ and hence may be computed by use of a restricted approach through the above mentioned dense set where $\alpha(t)$ is already known to be zero. By (5.2) $\alpha(t) = 0$ for $0 \leq t < 1$, and $\alpha(1) = 0$ by hypothesis 2 $(n = 0)$.

Corollary 5.1.

$$1. \quad \phi(t) \in L, \qquad 0 \leq t \leq 1;$$

$$2. \quad \int_0^1 \phi(t)t^n \, dt = 0, \qquad n = 0, 1, 2, \ldots$$

$$\Rightarrow \qquad \phi(t) = 0 \text{ almost everywhere}, \qquad 0 \leq t \leq 1.$$

If we set

$$\alpha(t) = \int_0^t \phi(y) \, dy,$$

then the continuous function $\alpha(t)$ vanishes identically by Theorem 5.1. But its derivative, also zero, is equal to $\phi(t)$ almost everywhere.

We may use Theorem 5.1 to improve Theorem 7.2 of Chapter 5. There we showed that the identical vanishing of a generating function implied the identical vanishing of the determining function. We can now draw the same conclusion if the generating function vanishes at a certain set of points in arithmetic progression. The result is usually called Lerch's theorem, named after M. Lerch [1903].

Theorem 5.2.

$$1. \quad f(s) = \int_0^\infty e^{-st} \, d\alpha(t), \qquad \sigma > \sigma_c;$$

$$2. \quad f(s_0 + nq) = 0, \qquad q > 0, n = 0, 1, 2, \ldots$$

$$\Rightarrow \qquad f(s) \equiv 0, \qquad \sigma > \sigma_c.$$

Since

$$(5.6) \qquad f(s) = s \int_0^\infty e^{-st}\alpha(t) \, dt, \qquad \sigma > \sigma_c, \ \sigma > 0,$$

and since hypothesis 2 is really unaltered if a finite number of the zeros of f at the beginning of the progression are omitted from consideration, we may assume that $\sigma_0 > 0$, $\sigma_0 > \sigma_c$. Then

$$(5.7) \qquad \int_0^\infty e^{-nt}\beta(t)\,dt = 0, \qquad n = 0, 1, 2, \ldots$$

$$\beta(t) = \exp\left(-\frac{s_0 t}{q}\right)\alpha\left(\frac{t}{q}\right).$$

But (5.7) implies

$$\int_0^1 t^n\beta\left(\log\frac{1}{t}\right)dt = 0, \qquad n = 0, 1, 2, \ldots,$$

and by Corollary 5.1 $\beta(t)$ is zero almost everywhere. The same is true of $\alpha(t)$, and from (5.6) we have $f(s) \equiv 0$, as desired.

Even though Theorem 5.2 is valid if the progression starts at an arbitrary point of the region of convergence of the transform (however far to the right) it must not be supposed that Theorem 5.1 is valid if the sequence of integers n of hypothesis 2 is truncated at its beginning. Thus if $\alpha(t) = U(t)$, the unit function, all moments μ_n are zero except μ_0. But $U(t) \not\equiv 0$. In this connection we call attention to a generalization due to C. Muntz [1914]:

> Theorem 5.2 remains true if the sequence of exponents n of hypothesis 2 is replaced by another sequence p_n (not necessarily integers) such that

$$(5.8) \qquad p_0 = 0, \qquad \lim_{n\to\infty} p_n = \infty, \qquad \sum_{n=1}^\infty \frac{1}{p_n} = \infty.$$

Also the Weierstrass approximation theorem can be similarly improved as follows: *Any $f(x) \in C$ on $a \leq x \leq b$ can be uniformly approximated there arbitrarily closely by a finite linear combination of the functions x^{p_n}, equations (5.8) holding.*

6. Hausdorff's Moment Theorem

We now prove Hausdorff's [1921] main result. Define the difference operator $\triangle$ as

$$\triangle\mu_n = \mu_{n+1} - \mu_n,$$

and introduce the following terminology.

Definition 6.1.

A sequence $\{\mu_n\}_0^\infty$ is completely monotonic (c.m.)

$$\Leftrightarrow \qquad (-1)^k \Delta^k \mu_n \geqq 0, \qquad k, n = 0, 1, 2, \ldots$$

Examples of such sequences are

$$1, 1, 1, \ldots$$

$$1, 0, 0, \ldots$$

(6.1) $$1, a, a^2, \ldots \qquad 0 \leqq a \leqq 1.$$

(6.2) $$1, \frac{1}{2}, \frac{1}{3}, \ldots$$

But $\left\{\dfrac{1}{n!}\right\}_0^\infty \notin$ c.m. For, $\Delta^2 \mu_0 = -\dfrac{1}{2}$. Also $\left\{\dfrac{1}{(n+1)!}\right\}_0^\infty \notin$ c.m.

Hausdorff's theorem now reads as follows.

Theorem 6.

(6.3) $$1. \quad \mu_n = \int_0^1 t^n \, d\alpha(t), \qquad n = 0, 1, 2, \ldots;$$

$$2. \quad \alpha(t) \in \uparrow, \qquad 0 \leqq t \leqq 1$$

(6.4) $$\Leftrightarrow \qquad \{\mu_n\}_0^\infty \in \text{c.m.}$$

For example, the sequences (6.1) and (6.2) are

$$a^n = \int_0^1 t^n \, dU(t - a),$$

$$\frac{1}{n+1} = \int_0^1 t^n \, dt,$$

and in both cases the integrator function is nondecreasing.

The necessity of condition (6.4) is immediate. For, from (6.3) we have

$$(-\Delta)^k \mu_n = \int_0^1 (1 - t)^k t^n \, d\alpha(t) \geqq 0.$$

To prove the sufficiency we establish first a series of lemmas, involving a "moment operator" M which we now define. It depends on a given sequence of constants $\{\mu_n\}_0^\infty$.

Definition 6.2. For any polynomial

$$P(x) = \sum_{k=0}^{n} a_k x^k,$$

$$M[P(x)] = \sum_{k=0}^{n} a_k \mu_k.$$

Note that M is defined so as to operate on polynomials only and that it is attached to a given sequence μ_n. Observe also that if it were known that the sequence had the representation (6.3) then

$$M[P] = \int_0^1 P(t)\, d\alpha(t).$$

Thus, after the theorem is proved, it will follow that M is a *positive* operator, carrying positive polynomials into positive numbers. It is clearly linear:

$$M[c_1 P_1(x) + c_2 P_2(x)] = c_1 M[P_1(x)] + c_2 M[P_2(x)].$$

Lemma 6.1.

1. $N[P(x)]$ is defined as in Definition 6.2 for M except that μ_k is replaced by μ_{k+1}.

$\Rightarrow$ $\qquad\qquad M[xP(x)] = N[P(x)].$

The proof is immediate.

Lemma 6.2.

(6.5) $\qquad\qquad M[(1-x)^m x^n] = (-\triangle)^m \mu_n.$

For, if E is the translation operator,

$$E[\mu_n] = \mu_{n+1},$$

then $-\triangle = 1 - E$ and

$$(-\triangle)^m = (1-E)^m = \sum_{k=0}^{m} (-1)^k \binom{m}{k} E^k.$$

Both sides of (6.5) are seen to be

$$\sum_{k=0}^{m} (-1)^k \binom{m}{k} \mu_{n+k}.$$

Lemma 6.3.

$$(6.6) \qquad \sum_{k=0}^{n} \binom{n}{k} (-\triangle)^{n-k} \mu_k = \mu_0.$$

For, by Lemma 6.2 this sum is

$$M\left[\sum_{k=0}^{n} \binom{n}{k} x^k (1-x)^{n-k} \right] = M[1] = \mu_0.$$

For example, if $n = 2$, (6.6) becomes

$$(\mu_0 - 2\mu_1 + \mu_2) + 2(\mu_1 - \mu_2) + \mu_2 = \mu_0.$$

We now define a special type of polynomial. It is the *Bernstein polynomial of degree n* for the function x^m. It is known that this polynomial tends uniformly on $0 \le x \le 1$ to x^m as the degree becomes infinite. This property provides the motivation of our proof of Theorem 6, but no use of the property will be made therein.

Definition 6.3.

$$F_{m,n}(x) = \sum_{k=0}^{n} \left(\frac{k}{n}\right)^m \binom{n}{k} x^k (1-x)^{n-k}.$$

Lemma 6.4. For $m,n = 1,2,3,\dots$

$$(6.7) \qquad F_{m,n}(x) = F_{m-1,n}(x) - \left(\frac{n-1}{n}\right)^{m-1} (1-x) F_{m-1,n-1}(x).$$

If we note that

$$\frac{k}{n}\binom{n}{k} = \binom{n}{k} - \binom{n-1}{k}$$

then $F_{m,n}(x) - x^n$ may be written as

$$\sum_{k=0}^{n-1} \left(\frac{k}{n}\right)^{m-1} \binom{n}{k} x^k (1-x)^{n-k} - \sum_{k=0}^{n-1} \left(\frac{k}{n}\right)^{m-1} \binom{n-1}{k} x^k (1-x)^{n-k}.$$

This first sum is $F_{m-1,n}(x) - x^n$, so we need only check that the second is $F_{m-1,n-1}(x)$ except for the factor $(1 - x)\,(n - 1)^{m-1}/n^{m-1}$. But this is immediate from Definition 6.3. For example, if $n = 2$ and $m = 3$, the lemma becomes

$$\left(\frac{1}{2}\right)^3 2x(1 - x) + x^2 = \left(\frac{1}{2}\right)^2 2x(1 - x) + x^2 - \left(\frac{1}{2}\right)^2 (1 - x)x.$$

Lemma 6.5.

$$(6.8) \qquad\qquad \lim_{n \to \infty} M[F_{m,n}(x)] = \mu_m.$$

This lemma contains the essence of the proof. For note that if we assumed the representation (6.3), equation (6.8) would be

$$\lim_{n \to \infty} \int_0^1 F_{m,n}(t)\, d\alpha(t) = \int_0^1 t^m\, d\alpha(t),$$

and this would follow from the known uniform convergence of the Bernstein polynomial $F_{m,n}$ to its defining function t^m. We prove the lemma by induction on m. For $m = 0$ the sequence on the left of (6.8) is μ_0 for *all* n by Lemma 6.3. Now assume (6.8) true for a fixed integer $m - 1$ but for *every* moment operator M (that is, corresponding to every sequence μ_n). We now prove (6.8) for m by applying the operator M to equation (6.7). On the right we obtain by Lemma 6.1.

$$M[F_{m-1,n}(x)] - \left(\frac{n - 1}{n}\right)^{m-1} M[F_{m-1,n-1}(x)]$$

$$+ \left(\frac{n - 1}{n}\right)^{m-1} N[F_{m-1,n-1}(x)],$$

and by the induction assumption this approaches

$$\mu_{m-1} - \mu_{m-1} + \mu_m,$$

as $n \to \infty$. Here we have used the fact that the sequence corresponding to N is derived from that for M by increasing each subscript by unity. This completes the proof.

Lemma 6.6.

$$1. \quad \alpha_n(x) \in \uparrow, \qquad n = 0, 1, \ldots, \quad 0 \le x \le 1;$$

$$2. \quad |\alpha_n(x)| \le A, \qquad n = 0, 1, \ldots, \quad \text{some } A$$

$\Rightarrow \quad$ There exists integers $n_0 < n_1 < n_2 < \cdots$ and $\alpha(x) \in \uparrow$,

$0 \le x \le 1$, such that

$$\lim_{k \to \infty} \alpha_{n_k}(x) = \alpha(x), \qquad 0 \le x \le 1.$$

This is E. Helly's theorem and will be assumed. See D. V. Widder [1946, p. 27].

We turn now to the proof of Theorem 6. By Lemma 6.2 equation (6.8) becomes

$$\lim_{n \to \infty} \sum_{k=0}^{n} \left(\frac{k}{n}\right)^m \binom{n}{k}(-\Delta)^{n-k}\mu_k = \mu_m,$$

and this sum may be written as the Stieltjes integral

$$\int_0^1 t^m \, d\alpha_n(t),$$

where $\alpha_n(t)$ is a step-function with jumps defined as follows

$$\alpha_n(0+) = \alpha_n(0) = 0,$$

$$\alpha_n\left(\frac{k}{n}+\right) - \alpha_n\left(\frac{k}{n}-\right) = \binom{n}{k}(-\Delta)^{n-k}\mu_k, \qquad k = 1, 2, \ldots, n - 1,$$

$$\alpha_n(1) - \alpha_n(1-) = \mu_n.$$

By Lemma 6.3 $\alpha_n(1) = \mu_0$ and by (6.4) $\alpha_n(t) \in \uparrow$. By Helly's theorem

$$\lim_{k \to \infty} \alpha_{n_k}(t) = \alpha(t)$$

for a suitable $\alpha(t) \in \uparrow$. Hence (6.8) becomes

$$\lim_{k \to \infty} \int_0^1 t^m \, d\alpha_{n_k}(t) = \mu_m.$$

If the limit and integral signs may be interchanged this becomes

$$(6.9) \qquad \int_0^1 t^m \, d\alpha(t) = \mu_m, \qquad m = 0, 1, 2, \ldots,$$

as was to be proved.

That the interchange is valid becomes evident if one integrates by parts and applies Lebesgue's bounded convergence theorem. Note that (6.9) is trivially true for $m = 0$ since $\alpha_n(1) = \alpha(1) = \mu_0$. This completes the proof.

7. Bernstein's Theorem

We now make use of Hausdorff's result, Theorem 6, to obtain a necessary and sufficient condition for the representation of a function as a Laplace–Stieltjes integral. The condition to be used is that the function should be *completely monotonic*, a property of functions analogous to that defined in §6 for sequences.

Definition 7.

A function $f(x)$ is completely monotonic $(f(x) \in \text{c.m.})$ $a < x < \infty$

$\Leftrightarrow$ A. $f(x) \in C^\infty$, $a < x < \infty$,

   B. $(-1)^k f^{(k)}(x) \geqq 0$, $k = 0, 1, 2, \ldots,$ $a < x < \infty$.

If, in addition, $f(a+) < \infty$, then $f(x) \in \text{c.m.}$ in $a \leqq x < \infty$.

As examples, the functions

$$\frac{1}{x}, \qquad \frac{1}{1 - e^{-x}}, \qquad \zeta(x + 1)$$

are all completely monotonic in $0 < x < \infty$ but not in $0 \leqq x < \infty$. On the other hand

$$\frac{1}{x + 1}, \qquad \sum_{k=1}^{\infty} \frac{e^{-kx}}{k^2}$$

are completely monotonic in $0 \leqq x < \infty$. But $(1 + x^2)^{-1} \notin \text{c.m.}$ on (a, ∞) for any a. This will be apparent from Theorem 7.

S. Bernstein's theorem [1928, p. 56] follows.

Theorem 7.

$$1. \quad f(x) \in \text{c.m.}, \qquad a \leq x < \infty$$

$$(7.1) \quad \Leftrightarrow \qquad f(x) = \int_0^\infty e^{-xt}\, d\alpha(t), \qquad a \leq x < \infty,$$

where $\alpha(t) \in \uparrow$. B on $0 \leq t < \infty$.

Since $f(x + a) \in \text{c.m.}$ on $0 \leq x < \infty$ under hypothesis 1 and since

$$\int_0^t e^{-ay}\, d\alpha(y) \in \uparrow$$

when $\alpha(t) \in \uparrow$, we may assume without loss of generality that $a = 0$. If equation (7.1) holds then

$$(-1)^k f^{(k)}(x) = \int_0^\infty e^{-xt} t^k\, d\alpha(t) \geq 0, \qquad 0 < x < \infty,$$

$$f(0+) = \int_0^\infty d\alpha(t) < \infty.$$

We have here used Theorems 4.1 and 4.2 of Chapter 5. Thus the necessity of hypothesis 1 is proved. To prove the sufficiency we use the following lemma.

Lemma 7.

$$1. \quad f(x) \in \text{c.m.}, \qquad 0 \leq x < \infty;$$

$$2. \quad \delta > 0$$

$$\Rightarrow \qquad \{f(n\delta)\}_0^\infty \in \text{c.m.}$$

Set $g(x) = f(\delta x)$. Then $g(x) \in \text{c.m.}, 0 \leq x < \infty$. Now note that

$$-\Delta g(x) = g(x) - g(x + 1) \in \text{c.m.}, \quad 0 \leq x < \infty.$$

For,

$$(-1)^k[g^{(k)}(x) - g^{(k)}(x + 1)] = (-1)^{k+1} g^{(k+1)}(X), \text{some } X, x < X < x+1,$$

by the mean-value theorem. The right-hand side is ≥ 0. Applying the result successively k times we see that $(-\Delta)^k g(x)$ is completely monotonic and hence ≥ 0 on $0 \leq x < \infty$ for $k = 1, 2, 3, \ldots$. In particular,

$$(-1)^k \Delta^k g(n) = (-1)^k \Delta^k f(n\delta) \geq 0, \qquad k, n = 0, 1, 2, \ldots,$$

as we wished to prove.

Now to prove the theorem choose δ of the lemma equal to $1/m$, the reciprocal of a positive integer. Hence $\left\{f\left(\dfrac{n}{m}\right)\right\}_{n=0}^{\infty} \in$ c.m. and by Hausdorff's theorem

$$(7.2) \qquad f\left(\frac{n}{m}\right) = \int_0^1 t^n \, d\alpha_m(t), \qquad n = 0, 1, 2, \ldots$$

for some normalized nondecreasing and bounded function $\alpha_m(t)$. That is,

$$f(n) = \int_0^1 t^{nm} \, d\alpha_m(t) = \int_0^1 t^n \, d\alpha_m(t^{1/m})$$

$$= \int_0^1 t^n \, d\alpha_1(t), \qquad (m = 1).$$

By Theorem 5.1 $\alpha_m(t^{1/m}) = \alpha_1(t)$. Thus equation (7.2) becomes

$$f\left(\frac{n}{m}\right) = \int_0^1 t^n \, d\alpha_1(t^m) = \int_0^1 t^{n/m} \, d\alpha_1(t), \qquad n = 0, 1, 2, \ldots$$

$$(7.3) \qquad = \int_{0+}^1 t^{n/m} \, d\alpha_1(t) = \int_0^{\infty} e^{-nt/m} \, d\alpha_1(e^{-t}), \qquad n = 1, 2, \ldots .$$

Set $\alpha_1(e^{-t}) = \alpha(t)$ and form the function

$$F(x) = \int_0^{\infty} e^{-xt} \, d\alpha(t).$$

Equation (7.3) shows that $F(x)$ and $f(x)$ coincide for all positive rational values of x. Since both are continuous on $0 \leq x < \infty$, they coincide there and $f(x)$ has the desired representation. It is of course evident that $\alpha(t)$, coming as it does from Hausdorff's representation, is bounded.

Corollary 7.

$$f(x) \in \text{c.m.}, \qquad a < x < \infty$$

$$\Leftrightarrow \qquad f(x) = \int_0^{\infty} e^{-xt} \, d\alpha(t), \qquad a < x < \infty,$$

where $\alpha(t) \in \uparrow$ on $0 \leqq t < \infty$.

Again taking $a = 0$, we note that for every $\delta > 0$ $f(x + \delta) \in$ c.m. on $0 \leq x < \infty$. By Bernstein's theorem

$$f(x + \delta) = \int_0^\infty e^{-xt} \, d\alpha_\delta(t), \qquad 0 \leq x < \infty,$$

where $\alpha_\delta(t) \in \uparrow$. B on $0 \leq t < \infty$. That is,

$$(7.4) \qquad f(x) = \int_0^\infty e^{-xt} \, d\alpha(t), \qquad x > \delta,$$

$$\alpha(t) = \int_0^t e^{\delta y} \, d\alpha_\delta(y).$$

By Theorem 5.2, the uniqueness of a Laplace representation, $\alpha(t)$ must be independent of δ in spite of appearances. Since δ was arbitrary (7.4) must hold for $x > 0$.

For the above examples, the function $\alpha(t)$ corresponding to the completely monotonic function $1/x$ is the unbounded function t; that corresponding to $1/(x + 1)$ is the bounded function $1 - e^{-t}$. Since

$$\frac{1}{x^2 + 1} = \int_0^\infty e^{-xt} \, d(1 - \cos t), \qquad 0 < x < \infty,$$

and since $1 - \cos t$ is not monotonic, it follows that $1/(x^2 + 1)$ cannot be completely monotonic in any neighborhood of $+\infty$.

8. Bounded Determining Function

An application of Bernstein's theorem enables us to obtain necessary and sufficient conditions for the representation of a function as a Laplace–Lebesgue integral with bounded determining function.

Theorem 8.

$$1. \quad |f^{(k)}(x)| \leq \frac{Mk!}{x^{k+1}}, \quad \text{some } M, \qquad 0 < x < \infty, \quad k = 0, 1, 2$$

$$(8.1) \quad \Leftrightarrow \qquad f(x) = \int_0^\infty e^{-xt} \phi(t) \, dt, \qquad |\phi(t)| \leq M.$$

Assuming first the representation (8.1), we have for $k = 0, 1, 2, \ldots$

$$|f^{(k)}(x)| \leqq \int_0^\infty e^{-xt} t^k |\phi(t)| \, dt \leqq \frac{Mk!}{x^{k+1}}, \qquad 0 < x < \infty.$$

Thus the necessity of conditions 1 is proved.

Conversely, from conditions 1 we see at once that

$$\frac{-Mk!}{x^{k+1}} \leqq (-1)^k f^{(k)}(x) \leqq \frac{Mk!}{x^{k+x}}, \qquad 0 < x < \infty.$$

This is equivalent to saying that $(M/x) - f(x)$ and $f(x) + (M/x)$ are completely monotonic in $0 < x < \infty$. By Corollary 7 there must exist nondecreasing functions $\beta(t)$ and $\gamma(t)$ such that

$$\frac{M}{x} - f(x) = \int_0^\infty e^{-xt} \, d\beta(t), \qquad 0 < x < \infty;$$

$$\frac{M}{x} + f(x) = \int_0^\infty e^{-xt} \, d\gamma(t), \qquad 0 < x < \infty.$$

By the uniqueness theorem

$$(8.2) \qquad\qquad f(x) = \int_0^\infty e^{-xt} \, d\alpha(t),$$

where

$$(8.3) \qquad\qquad \alpha(t) = Mt - \beta(t) = \gamma(t) - Mt.$$

Here we have considered M/x as the Laplace–Stieltjes transform of Mt. Equations (8.3) show that $\alpha(t)$ is of bounded variation and hence has a derivative $\phi(t)$ almost everywhere. But (8.3) also shows that for any $\delta > -t$

$$(8.4) \qquad\qquad -M \leqq \frac{\alpha(t + \delta) - \alpha(t)}{\delta} \leqq M, \qquad t \geqq 0.$$

Hence we may use Lebesgue's limit theorem to show that

$$(8.5) \qquad\qquad \lim_{\delta \to 0} \int_0^t \frac{\alpha(y + \delta) - \alpha(y)}{\delta} \, dy = \int_0^t \phi(y) \, dy.$$

A simple transformation enables us to express the left side of (8.5) as

$$(8.6) \qquad \frac{1}{\delta} \int_t^{t+\delta} \alpha(y)\, dy - \frac{1}{\delta} \int_0^{\delta} \alpha(y)\, dy.$$

But (8.4) shows that $\alpha(t) \in C$, $0 \leq t < \infty$, so that (8.6) tends to $\alpha(t) - \alpha(0) = \alpha(t)$ as $\delta \to 0$. Thus $\alpha(t)$ is absolutely continuous,

$$\alpha(t) = \int_0^t \phi(y)\, dy$$

and (8.2) takes the form (8.1). The boundedness of ϕ follows at once from (8.4), and the proof is complete.

Corollary 8.

$$1. \quad |f^{(k)}(x)| \leq \frac{Mk!}{(x-a)^{k+1}}, \qquad a < x < \infty, \quad k = 0, 1, 2, \ldots$$

$$\Leftrightarrow \quad f(x) = \int_0^{\infty} e^{-xt}\phi(t)\, dt, \qquad |\phi(t)| \leq Me^{at}.$$

This result follows from the theorem after a translation in the variable x.

As an example, take $f(x) = e^{-x}/x$. Then

$$(-1)^k \frac{f^{(k)}(x)x^{k+1}}{k!} = \frac{e^{-x}}{k!} \sum_{p=0}^{k} p! \binom{k}{p} x^{k-p},$$

$$\left| \frac{f^{(k)}(x)x^{k+1}}{k!} \right| \leq e^{-x} \sum_{p=0}^{k} \frac{x^{k-p}}{(k-p)!} \leq e^{-x} e^{x} = 1.$$

Thus $f(x)$ must have the representation (8.1) with $M = 1$. Clearly

$$\frac{e^{-x}}{x} = \int_1^{\infty} e^{-xt}\, dt, \qquad 0 < x < \infty,$$

so that $\phi(t) = 0$ on $(0, 1)$ and $\phi(t) = 1$, on $(1, \infty)$ in accord with the theorem.

As a second example, consider the equation

$$1 = \int_0^{\infty} e^{-xt}\, dU(t).$$

Since $U(t)$ is not absolutely continuous the hypotheses 1 of Theorem 8 or of Corollary 8 must be violated. They clearly are, for $k = 0$, whatever the choice of M.

9. An Application of Bernstein's Thoerem

As an amusing consequence of Bernstein's theorem let us prove the following result of Y. Tagamlitzki [1946, p. 940].

Theorem 9.

$$1. \quad |f^{(k)}(x)| \leqq e^{-x}, \qquad 0 \leqq x < \infty, \quad k = 0, 1, 2, \dots$$

$$\Rightarrow \qquad f(x) = Ae^{-x} \qquad \text{for some } A, \quad |A| \leqq 1.$$

The hypothesis is equivalent to

$$-(-1)^k(e^{-x})^{(k)} \leqq (-1)^k f^{(k)}(x) \leqq (-1)^k(e^{-x})^{(k)},$$

so that the functions $e^{-x} \pm f(x)$ are completely monotonic on $0 < x < \infty$. Hence by Bernstein's theorem

$$e^{-x} - f(x) = \int_0^\infty e^{-xt}\, d\beta(t), \qquad \beta(t) \in \uparrow \quad (0, \infty),$$

$$e^{-x} + f(x) = \int_0^\infty e^{-xt}\, d\gamma(t), \qquad \gamma(t) \in \uparrow \quad (0, \infty).$$

That is,

$$f(x) = \int_0^\infty e^{-xt}\, d[U(t-1) - \beta(t)] = \int_0^\infty e^{-xt}\, d[\gamma(t) - U(t-1)].$$

By the uniqueness theorem

$$U(t-1) - \beta(t) = \gamma(t) - U(t-1),$$

$$(9.1) \qquad 2\Delta U(t-1) = \Delta\beta(t) + \Delta\gamma(t),$$

where Δ is the difference operator, $\Delta f(t) = f(t + \delta) - f(t)$. Since each term on the right is $\geqq 0$ when $\delta > 0$, it follows that each must be zero when the left is zero. Since $U(t-1)$ is constant except for a single jump at $t = 1$, the functions $\beta(t)$ and $\gamma(t)$ must have the same property.

Thus $f(x) = Ae^{-x}$, where A is the amount of the jump of $U(t-1)$ $- \beta(t)$ at $t = 1$. That $|A| \leq 1$ now follows from hypothesis 1 with $k = 0$. This completes the proof.

10. Completely Convex Functions

The hypotheses of Bernstein's theorem involve only the signs of the successive derivatives of a function. We now prove another result of the same kind.

Definition 10.

$$f(x) \text{ is completely convex on } a < x < b$$

$\Leftrightarrow$ A. $f(x) \in C^\infty, \quad a < x < b;$

         B. $(-1)^k f^{(2k)}(x) \geqq 0, \quad a < x < b.$

For example, $\sin x$ is completely convex on $(0, \pi)$, $\cos x$ on $(-\pi/2, \pi/2)$. We shall prove that any such function is entire, no matter how small the defining interval (a, b) may be. More precisely, $f(x)$ can be extended analytically into the complex plane so as to be entire. We prove this by use of a series of lemmas. In all of these we assume without further statement that $f(x) \in C^\infty$ in $a \leqq x \leqq b$, though less continuity would obviously be needed in some of them.

Lemma 10.1.

1. $|f^{(k)}(x)| \leqq M, \quad a \leqq x \leqq b, \quad \text{some } M, \quad k = 0, 1, 2, \ldots$

$\Rightarrow$ $f(x)$ is entire.

For, if $c = (a+b)/2$ we have by Taylor's series with remainder, after applying hypothesis 1 to be the remainder, that

$$\left| f(x) - \sum_{k=0}^{n} f^{(k)}(c) \frac{(x-c)^k}{k!} \right| \leqq M \frac{|x-c|^{n+1}}{(n+1)!} = o(1), \quad n \to \infty.$$

Hence

$$f(x) = \sum_{k=0}^{\infty} f^{(k)}(c) \frac{(x-c)^k}{k!}, \quad a \leqq x \leqq b.$$

But the series on the right converges for all x by hypothesis 1 and hence provides the desired analytic continuation.

Lemma 10.2.

$$1. \quad M_k = \max_{a \leq x \leq b} |f(x)|, \qquad k = 0, 1, 2$$

$$\Rightarrow \qquad M_1 \leq \frac{2M_0}{b-a} + \frac{M_2}{2}(b-a).$$

Let c be a point where $|f'(x)|$ attains its maximum. By Taylor's theorem (see Exercise 15 of this chapter)

$$f(b) - f(c) = (b-c)f'(c) + \tfrac{1}{2}f''(X)(b-c)^2, \qquad c < X < b,$$

$$f(a) - f(c) = (a-c)f'(c) + \tfrac{1}{2}f''(Y)(a-c)^2, \qquad a < Y < c.$$

Subtracting these equations, we have

$$(b-a)M_1 \leq |f(b) - f(a)| + \frac{1}{2}|f''(X)|(b-c)^2 + \frac{1}{2}|f''(Y)|(a-c)^2,$$

$$(b-a)M_1 \leq 2M_0 + \frac{1}{2}M_2[(b-c)^2 + (a-c)^2] \leq 2M_0 + \frac{1}{2}M_2(b-a)^2.$$

We have here used the fact that the bracket, considered as a function of c, takes its maximum at an end point of (a, b). This result is due to J. Hadamard. For a reference and the above proof see T. Carleman [1926, p.11].

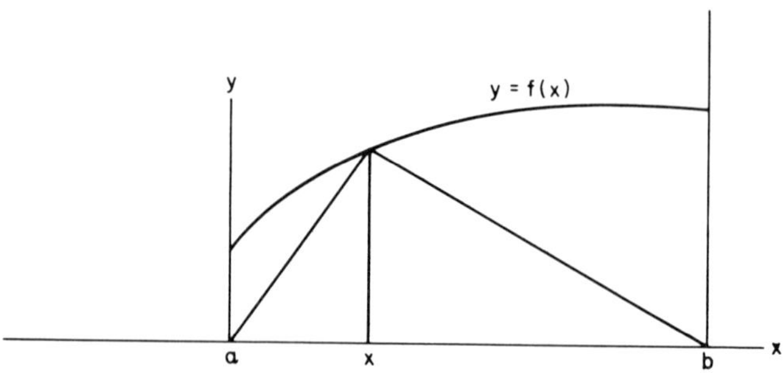

Figure 1

Lemma 10.3.

$$1. \quad f(x) \geqq 0, \qquad a \leqq x \leqq b;$$
$$2. \quad f''(x) \leqq 0, \qquad a \leqq x \leqq b$$

$$\Rightarrow \qquad f(x) \leqq \frac{2}{b-a} \int_a^b f(x)\, dx.$$

It is obvious graphically, from the convexity of the curve $y = f(x)$, that the triangle of the figure lies below the curve. By a comparison of areas we have the desired result.

Theorem 10.

$$1. \quad f(x) \text{ is completely convex in } a < x < b$$

$$\Rightarrow \qquad f(x) \text{ is entire.}$$

It is no restriction to take $a = 0$, $b = \pi$. Then

$$\int_0^\pi f(x) \sin x\, dx = f(\pi) + f(0) - \int_0^\pi f''(x) \sin x\, dx,$$

$$(10.1) \qquad -\int_0^\pi f''(x) \sin x\, dx \leqq \int_0^\pi f(x) \sin x\, dx.$$

Since $(-1)^k f^{(2k)}(x)$ is completely convex we may apply (10.1) to obtain

$$(-1)^k \int_0^\pi f^{(2k)}(x) \sin x\, dx \leqq (-1)^{k-1} \int_0^\pi f^{(2k-2)}(x) \sin x\, dx$$

$$\leqq \int_0^\pi f(x) \sin x\, dx = A.$$

Let δ be a small positive number and apply Lemma 10.3 with $a = \delta$, $b = \pi - \delta$,

$$0 \leqq (-1)^k \int_\delta^{\pi-\delta} f^{(2k)}(x) \sin x\, dx \leqq A/\sin \delta,$$

$$0 \leqq (-1)^k f^{(2k)}(x) \leqq \frac{2A}{(\pi - 2\delta) \sin \delta}, \qquad \delta \leqq x \leqq \pi - \delta.$$

That is, the even derivatives of $f(x)$ satisfy the hypothesis of Lemma 10.1. That the odd ones do also follows from Lemma 10.2, and the proof is complete. This result was originally due to the author [1940], but the present proof is due to R. P. Boas.

11. Summary

The principal results of the present chapter involve sereval definitions and analytic formulas. We summarize them as follows:

A.   The Laplace method.

$$\int_a^b e^{nh(x)}\phi(x)\,dx \sim \sqrt{2\pi}\,\phi(c)e^{nh(c)}(-nh''(c))^{-1/2}, \qquad n\to\infty,$$

if $h(x)$ has a single maximum at c, $a < c < b$, where $h''(c) < 0$.

B.   The Laplace inversion operator.

$$L_{k,t}[f] = \frac{(-1)^k}{k!}\left(\frac{k}{t}\right)^{k+1} f^{(k)}\left(\frac{k}{t}\right),$$

$$L_{k,t}\left[\int_0^\infty e^{-xt}\phi(t)\,dt\right] \to \phi(t), \qquad k\to\infty.$$

C.   The Stieltjes inversion operator.

$$M_{k,t}[f] = \frac{(-1)^k}{k!k!}\,[t^{2k+1}f^{(k+1)}(t)]^{(k)},$$

$$M_{k,t}\left[\int_0^\infty \frac{\phi(t)}{x+t}\,dt\right] \to \phi(t), \qquad k\to\infty.$$

D.   Hausdorff's theorem (completely monotonic sequences).

$$(-1)^k\Delta^k\mu_n \geq 0 \Leftrightarrow \mu_n = \int_0^1 t^n\,d\alpha(t), \qquad \alpha(t)\in\uparrow.$$

E.   Bernstein's theorem (completely monotonic functions).

$$(-1)^k f^{(k)}(x) \geq 0 \Leftrightarrow f(x) = \int_0^\infty e^{-xt}\,d\alpha(t), \qquad \alpha(t)\in\uparrow.$$

F.   Completely convex functions

$$(-1)^k f^{(2k)}(x) \geq 0 \Rightarrow f(x) \text{ is entire.}$$

EXERCISES

1.
$$\int_0^1 (1 - t^2)^n \, dt \sim \, ? \qquad n \to \infty.$$

Check by expressing the integral in terms of the Gamma function and using Stirling's formula.

Ans. $\tfrac{1}{2}(\pi/n)^{1/2}$.

2.
$$\int_0^1 (3 - e^t)^n \, dt \sim \, ? \qquad n \to \infty.$$

Ans. $2^{n+1}/n$.

3.
$$\int_0^\pi (\sin t)^{2n} \cos 2t \, dt \sim \, ? \qquad n \to \infty.$$

Ans. $-(\pi/n)^{1/2}$.

4.
$$\int_0^\pi (1 - \sin t)^n \, dt \sim \, ? \qquad n \to \infty.$$

Ans. $2/n$.

5. If $h''(x) < 0$ in $a \leq x \leq b$, $h'(c) = 0$ for $a < c < b$, $\phi(x) \in C$ in $a \leq x \leq b$ except at c where $\phi(c+)$ and $\phi(c-)$ exist, apply the Laplace asymptotic method to

$$\int_a^b e^{nh(x)}\phi(x) \, dx.$$

Ans. $\sqrt{\pi} \, \dfrac{[\phi(c+) + \phi(c-)]e^{nh(c)}}{\sqrt{-2nh''(c)}}.$

6. If $[x]$ means "greatest integer $\leq x$," prove

$$\lim_{k \to \infty} \frac{k^{k+1}}{k!} \int_0^2 e^{-kt}[t + 2]t^k \, dt = \frac{5}{2}.$$

7. If $\alpha(t) \in V^*$ in $0 \leq t \leq R$ for every R and if

$$f(x) = \int_0^\infty e^{-xt} \, d\alpha(t), \qquad x > 0,$$

prove

$$\lim_{k \to \infty} L_{k,t}\left[\frac{f(x)}{x}\right] = \alpha(t), \qquad 0 < t < \infty.$$

8.
$$\int_0^\pi e^{n(\sin x - x)}x^4 \, dx \sim \ ? \qquad n \to \infty.$$

$$\text{Ans.} \ \frac{6^{5/3}\Gamma(5/3)}{3n^{5/3}}.$$

9. Show that the sequence $1/(n+1)!$, $n = 0, 1, 2, \ldots$ is not completely monotonic.

10. Show that $1/(x^3 + x)$ is completely monotonic on $0 < x < \infty$.

11. Prove

$$\lim_{k \to \infty} L_{k,t}[\sqrt{x}] = \frac{-1}{2\sqrt{\pi}\, t^{3/2}}.$$

12. If

$$f(x) = \int_0^\infty (e^{-xt} - 1)t^{-3/2} \, dt$$

compute by use of formula (2.10)

$$\lim_{k \to \infty} L_{k,t}[f(x)].$$

13. For $a > 0$ and k a positive integer prove that

$$\max_{a \leq x \leq \infty} \frac{(-1)^k}{k!}(x - a)^{k+1}\frac{d^k}{dt^k}x^{-2} = \frac{1}{a}\left(\frac{k+1}{k+2}\right)^{k+2} < \frac{1}{ae}.$$

14. For $a > 0$ prove that

$$aet \leqq e^{at}, \qquad 0 < t < \infty.$$

Use this result and Exercise 13 to provide an illustration for Corollary 8.

15. In the proof of Lemma 10.2 we tacitly assumed $a < c < b$. If that assumption was not justified, complete the proof of the lemma.

16. If for $\lambda > 0$

$$|f^{(k)}(x)| \leqq \lambda^k e^{-\lambda x}, \qquad 0 < x < \infty, \quad k = 0, 1, 2, \ldots,$$

prove $f(x) = Ae^{-\lambda x}$, where $|A| \leqq 1$.

17. If for $a > 0, b > 0$

$$|f^{(k)}(x)| \leqq ae^{-x} + b2^k e^{-2x}, \qquad 0 < x < \infty, \quad k = 0, 1, 2, \ldots,$$

prove that $f(x) = Ae^{-x} + Be^{-2x}$, where $|A| \leqq a, |B| \leqq b$.

18. If $g(x)$ is the completely monotonic sum of a Dirichlet series convergent for $x > 0$ and if

$$|f^{(k)}(x)| \leqq g^{(k)}(x), \qquad 0 < x < \infty, \quad k = 0, 1, 2, \ldots,$$

prove that $f(x)$ is also the sum of a Dirichlet series.

7 **_The Convolution Transform_**

1. Introduction

The Laplace and Stieltjes transforms both take the following form, as we shall see, after exponential changes of variable,

$$(1.1) \qquad f(x) = \int_{-\infty}^{\infty} G(x - t)\phi(t)\, dt = G * \phi.$$

This is called the convolution transform. Since we were able to invert both of those special cases of (1.1) by linear differential operators it is natural to seek an analogous inversion of (1.1) generally. In each special case it was possible to choose the operator in such a way as to produce a modified kernel G_n which took the form of a power g^n, so that the Laplace asymptotic method was available. For more general kernels G we abandon this special tool but retain many of the features of the method. For a large class of kernels G we shall be able to discover a linear differential operator of order n, with constant coefficients, which when applied to f will produce a kernel G_n which is everywhere positive and which tends to the Dirac δ-function as n becomes infinite, just as the powers g^n did in the two special cases. We turn now to the details.

169

2. Definitions and Examples

We introduce the transform to be studied by the following definition.

Definition 2. The function $f(x)$ is a convolution transform of $\phi(t)$ if

$$(2.1) \qquad f(x) = G(x) * \phi(x) = \int_{-\infty}^{\infty} G(x - t)\phi(t)\, dt,$$

the integral converging for some x. The function $G(x)$ is called the kernel of the transform, ϕ is the determining, and f the generating function.

We illustrate the definition by a number of examples.

Example A. The unilateral Laplace transform

$$(2.2) \qquad F(x) = \int_{0}^{\infty} e^{-xt}\, \Phi(t)\, dt.$$

This takes the form (2.1) if

$$f(x) = e^x F(e^x),$$
$$G(x) = e^x \exp(-e^x),$$
$$\phi(t) = \Phi(e^{-t}).$$

Example B. The Stieltjes transform

$$F(x) = \int_{0}^{\infty} \frac{\Phi(t)}{x + t}.$$

Here

$$f(x) = F(e^x), \qquad G(x) = \frac{1}{e^x + 1}, \qquad \phi(t) = \Phi(e^t).$$

Example C.

$$f(x) = e^x \int_{x}^{\infty} e^{-t}\phi(t)\, dt.$$

This transform already has the form (2.1) without any change of variable if $G(t) = e^t\ (-\infty, 0)$, $G(t) = 0\ (0, \infty)$.

Example D.

$$F(x) = \frac{2}{\pi} \int_0^\infty \frac{t}{x^2 + t^2}\, \Phi(t)\, dt.$$

This might be called the *potential transform*, since it is clearly related to the Poisson integral representation of a function which is harmonic in a half-plane. We again need an exponential change of variable to attain the convolution form:

$$f(x) = F(e^{-x}), \qquad G(x) = \frac{2}{\pi}\frac{1}{1 + e^{-2x}}, \qquad \phi(t) = \Phi(e^{-t}).$$

3. Operational Calculus

To predict an inversion formula for the convolution transform it is useful to apply the notions of operational calculus. In it we treat some symbol, such as D for differentiation, as if it were a number throughout a series of calculations and then at the end we give it its original meaning. The justification of such a procedure lies in the algebra of the operator used. We shall not be concerned with the justification of the calculus but use it heuristically only.

Definition 3.1.

$$e^{aD}f(x) = f(x + a).$$

If D were a number we should have

$$e^{aD} = \sum_{k=0}^\infty \frac{a^k D^k}{k!}.$$

If D^k is now interpreted as a derivative, we have

$$e^{aD} f(x) = \sum_{k=0}^\infty \frac{a^k f^{(k)}(x)}{k!} = f(x + a)$$

by Maclaurin's series, at least if $f(x)$ is analytic. But note that the definition is taken to apply to any function.

We can use Definition 3.1 to obtain the interpretation as an operator of any function of D which has a Laplace integral representation. For example, if

$$\frac{1}{E(s)} = \int_{-\infty}^{\infty} e^{-st} G(t)\, dt,$$

then a suitable interpretation of $1/E(D)$ should be

$$\frac{1}{E(D)} \phi(x) = \left(\int_{-\infty}^{\infty} e^{-tD} G(t)\, dt \right) \phi(x)$$

$$= \int_{-\infty}^{\infty} G(t)[e^{-tD}\phi(x)]\, dt = \int_{-\infty}^{\infty} G(t)\phi(x - t)\, dt.$$

That is,

$$\frac{1}{E(D)} \phi(x) = G(x) * \phi(x),$$

$$(3.1) \qquad \phi(x) = E(D)\,[G(x) * \phi(x)].$$

We are thus led to a symbolic inversion of the convolution transform. On account of equation (3.1) we make the following definition.

Definition 3.2. The reciprocal of the bilateral Laplace transform of the kernel of a convolution transform is called the inversion function of the convolution transform.

Let us illustrate by Example C above. Here

$$(3.2) \qquad \frac{1}{E(s)} = \int_{-\infty}^{0} e^{-st} e^{t}\, dt = \frac{1}{1 - s},$$

so that (3.1) becomes

$$(3.3) \qquad (1 - D)e^{x} \int_{x}^{\infty} e^{-t}\phi(t)\, dt = \phi(x).$$

But this may be checked directly by differentiation if $\phi(x) \in C$, for example. Indeed if $\phi(x) \in L$ in $0 \le x \le R$ for every $R > 0$ and if the integral (3.3) converges for $x > 0$, then equation (3.3) holds for almost all positive x. This conclusion is typical of the general case, though we shall not here attempt proofs for such generality.

4. The Laguerre–Pólya Class

The conclusion suggested by the heuristic developments of the previous section can be proved rigorously when the inversion function $E(s)$ belongs to a large class of entire functions studied earlier by E. Laguerre [1882, p. 828] and G. Pólya [1913, p. 279] in another connection. We shall call the class E.

Definition 4.1. An entire function $E(s)$ belongs to class E

$$(4.1) \quad \Leftrightarrow \quad E(s) = A \exp(bs - cs^2)\, s^n \prod_{k=1}^{\infty} \left(1 - \frac{s}{a_k}\right) \exp(s/a_k),$$

where A, b, c, a_k are real, $c \geq 0$, n is a nonnegative integer, and

$$(4.2) \qquad \sum_{k=1}^{\infty} \frac{1}{a_k^{\,2}} < \infty.$$

We shall assume known the elements of the Weierstrass theory of primary factors. The convergence of the series (4.2) guarantees that the product (4.1) should converge and represent an entire function.

Laguerre proved, and Pólya added a refinement, that a sequence of polynomials, having real roots only, which converge uniformly in every compact set of the complex s-plane, approaches a function of class E. Conversely, every function of class E is the uniform limit of such a sequence. For example,

$$\exp(-s^2) = \lim_{k \to \infty} \left(1 - \frac{s^2}{k^2}\right)^{k^2},$$

and the polynomials $(1 - s^2/k^2)^{k^2}$ have real roots only. This example helps one see why the constant c of the definition must be ≥ 0.

In the definition we do not assume that the a_k are distinct, as indeed they were not in the above example. Moreover, we wish to include the case in which the product is a finite one. We may assume, without loss of generality, that the roots a_k are arranged in an order of increasing absolute value,

$$0 < |a_1| \leq |a_2| \leq |a_3| \leq \cdots.$$

Examples of functions of E are

$$(4.3) \qquad 1, \quad (1-s), \quad \exp(-s^2), \quad \cos s, \quad \frac{\sin s}{s}, \quad \frac{1}{\Gamma(1-s)}, \quad \frac{1}{\Gamma(s)}.$$

We also introduce a notation E_0 for a subset of E as follows.

Definition 4.2.

$$E(s) \in E_0$$

$$(4.4) \quad \Leftrightarrow \qquad E(s) = e^{bs} \prod_{k=1}^{\infty} \left(1 - \frac{s}{a_k}\right) \exp(s/a_k).$$

where b and a_k are real and finite, and

$$\sum_{k=1}^{\infty} \frac{1}{a_k^2} < \infty.$$

We have thus omitted the cases $c > 0$ and those for which $E(s)$ has only a finite number of roots. Moreover, now $E(0) = 1$. Thus the first three functions (4.3) and the last are not members of E_0; the others are.

Even though the whole class E is composed of functions that can be inversion functions of convolution transforms we restrict attention here to the smaller class E_0 for brevity of exposition. It is not the property of the class, originally used by Laguerre to define it, that we shall need, but other properties which we shall now develop.

5. Some Statistical Terms

Although we shall make no use of any facts from the theory of statistics it is advantageous to employ a few terms therefrom. A *frequency function* $p(x)$ is any function ≥ 0 on $-\infty < x < \infty$ for which

$$\int_{-\infty}^{\infty} p(x)\, dx = 1.$$

The *mean* of a frequency function $p(x)$ is

$$M_p = \int_{-\infty}^{\infty} xp(x)\, dx$$

when the number is finite.

The *variance* of a frequency function $p(x)$ of mean M_p is

$$V_p = \int_{-\infty}^{\infty} (x - M_p)^2 \, p(x) \, dx.$$

For example, if

$$p(x) = k(x, t) = \frac{\exp(-x^2/4t)}{(4\pi t)^{1/2}}$$

then $p(x)$ is a frequency function for each positive value of the parameter t with mean 0 and variance $2t$. These facts are easily derived from
formula G, §2 of Chapter 5,

$$(4.5) \qquad \exp(ts^2) = \int_{-\infty}^{\infty} e^{-sy} k(y, t) \, dy, \qquad t > 0.$$

It is sometimes helpful to think of $p(x)$ as defining the manner in
which a unit mass is spread over the x axis. Then M_p is the center of
gravity of this linear bar and V_p is its moment of inertia about the center
of gravity. The more the mass is concentrated near this "fulcrum" the
smaller will be the bar's moment of inertia. Thus the variance measures
the tendency of the mass to lie away from or to vary from the mean.

6. Properties of the Laguerre–Pólya Kernels

We shall call any function $G(x)$ which has a bilateral Laplace
transform whose reciprocal lies in E a *Laguerre–Pólya kernel*. We now
develop some of the properties of such kernels.

Theorem 6.1.

$$E(s) \in E_0$$

$$\Rightarrow \qquad |E(\sigma + iy)| \geqq |E(\sigma)|, \qquad -\infty < \sigma < \infty.$$

For, if $E(s)$ is defined by (4.4), then

$$|E(\sigma + iy)| = e^{b\sigma} \prod_{k=1}^{\infty} \left[\left(1 - \frac{\sigma}{a_k}\right)^2 + \frac{y^2}{a_k^2}\right]^{1/2} \exp\left(\frac{\sigma}{a_k}\right)$$

$$\geqq e^{b\sigma} \prod_{k=1}^{\infty} \left|1 - \frac{\sigma}{a_k}\right| \exp\left(\frac{\sigma}{a_k}\right) = |E(\sigma)|.$$

Theorem 6.2.

$$1. \quad E(s) \in E_0;$$

$$2. \quad R > 0, \qquad p > 0$$

$$(6.1) \quad \Rightarrow \qquad \frac{1}{E(\sigma + iy)} = O\left(\frac{1}{|y|^p}\right), \qquad |y| \to \infty$$

uniformly in $|\sigma| \leqq R$.

That is, $1/E(s)$ is uniformly small in any vertical strip of the complex s-plane, smaller than any power of $1/|y|$. This theorem would not be true for functions of the whole class E, as one sees by observing the first two functions (4.3). To prove it choose $N > p$ and so large that $|a_k| \geqq R$ when $k > N$. This is possible since $|a_k| \to \infty$ as $k \to \infty$. Set

$$E_N(s) = \prod_{k=N+1}^{\infty} \left(1 - \frac{s}{a_k}\right)\exp\left(\frac{s}{a_k}\right).$$

By Theorem 6.1

$$|E(\sigma + iy)| \geqq e^{b\sigma}|E_N(\sigma)| \prod_{k=1}^{N} \frac{|y|}{|a_k|} \exp\left(\frac{-R}{|a_k|}\right) \qquad |\sigma| \leqq R.$$

Since $e^{b\sigma}|E_N(\sigma)|$ has no roots and is continuous in $|\sigma| \leqq R$ it has a positive lower bound there. Hence for a suitable constant M

$$\left|\frac{1}{E(\sigma + iy)}\right| \leqq \frac{M}{|y|^N}, \qquad |\sigma| \leqq R.$$

This proves the result.

Theorem 6.3.

$$1. \quad E(s) \in E_0$$

$$(6.2) \quad \Rightarrow \qquad \frac{1}{E(s)} = \int_{-\infty}^{\infty} e^{-st} G(t) \, dt. \qquad |\sigma| < |a_1|$$

for some function $G(t)$. Here a_1 is a root of $E(s)$ nearest the origin.

That is, the reciprocal of every function $E(s)$ of E_0 is a bilateral Laplace transform which converges in a strip symmetric about the imaginary axis and extending out to the nearest zero of $E(s)$. To prove this we have only to apply Theorem 11.1, Chapter 5. By Theorem 6.2 with $p = 2$ we see that $1/E(\sigma + iy) \in L$, $-\infty < y < \infty$ for each σ in $|\sigma| < |a_1|$ and that $1/E(\sigma + iy)$ tends uniformly to zero in that strip. It is clearly analytic there, so that all conditions of Hamburger's theorem are satisfied.

We shall now prove that the function $G(t)$ of Theorem 6.3 is a frequency function and compute its mean and variance.

Theorem 6.4.

$$1. \quad E(s) \in E_0, \qquad E(s) = e^{bs} \prod_{k=1}^{\infty} \left(1 - \frac{s}{a_k}\right) \exp\left(\frac{s}{a_k}\right);$$

$$(6.3) \qquad 2. \quad G(t) = \frac{1}{2\pi i} \int_{-i\infty}^{i\infty} \frac{e^{st}}{E(s)} \, ds, \qquad -\infty < t < \infty$$

$\Rightarrow$ A. $G(t)$ is a frequency function;

       B. $M_G = b$;

$$C. \quad V_G = \sum_{k=1}^{\infty} \frac{1}{a_k^2}.$$

In defining $G(t)$ by equation (6.3) we have inverted the integral (6.2) by Theorem 7.3 of Chapter 5. It is clear from Theorem 6.2 that the integral (6.3) converges absolutely for all t. Indeed, it defines $G(t)$ as a function of class C^∞ by virtue of the relations (6.1). To prove conclusion B we differentiate equation (6.2) with respect to s and set $s = 0$,

$$\frac{E'(0)}{E^2(0)} = E'(0) = \int_{-\infty}^{\infty} tG(t) \, dt.$$

But from the definition of $E(s)$ we have by logarithmic differentiation

$$(6.4) \qquad \frac{E'(s)}{E(s)} = b + \sum_{k=1}^{\infty} \left(\frac{1}{s - a_k} + \frac{1}{a_k}\right).$$

The validity of this step and the convergence of the series (6.4) derive from the general Weierstrass theory which we are assuming. Setting $s = 0$ in (6.4) gives B.

Set $F(s) = e^{bs}/E(s)$ and compute $F''(0)$ from the integral (6.2),

$$F''(0) = \int_{-\infty}^{\infty} (t - b)^2 G(t) \, dt.$$

On the other hand

$$\frac{F'(s)}{F(s)} = -\sum_{k=1}^{\infty}\left(\frac{1}{s-a_k} + \frac{1}{a_k}\right),$$

again by logarithmic differentiation. In particular, $F'(0) = 0$. A second differentiation gives

$$\frac{FF'' - (F')^2}{F^2} = \sum_{k=1}^{\infty} \frac{1}{(s-a_k)^2}.$$

For $s = 0$

$$F''(0) = \sum_{k=1}^{\infty} \frac{1}{a_k{}^2},$$

and conclusion C is established.

Finally, to prove A set $s = 0$ in (6.2), obtaining

$$1 = \int_{-\infty}^{\infty} G(t)\, dt.$$

It remains only to show that $G(t) \geqq 0$, and we shall do this by a limit process. We set

$$P_n(s) = e^{bs} \prod_{k=1}^{n}\left(1 - \frac{s}{a_k}\right)\exp\left(\frac{s}{a_k}\right)$$

and show first that

$$(6.5) \qquad \frac{1}{P_n(s)} = \int_{-\infty}^{\infty} e^{-st}g_n(t)\, dt,$$

where $g_n(t)$ is a frequency function. If $g_n(t) \to G(t)$ when $n \to \infty$, as it is reasonable to expect since $P_n(s) \to E(s)$, then $G(t)$ will also be a frequency function.

Clearly

$$\frac{1}{1-s} = \int_{-\infty}^{\infty} e^{-st}g(t)\, dt, \qquad \sigma < 1$$

if $g(t)$ is defined as e^t on $(-\infty, 0)$ and as zero elsewhere. Then

$$\frac{1}{\left(1 - \dfrac{s}{a_k}\right)\exp\left(\dfrac{s}{a_k}\right)} = |a_k| \int_{-\infty}^{\infty} e^{-st}g(a_k t - 1)\, dt, \qquad |\sigma| < |a_k|.$$

By use of Theorem 8 of Chapter 5 we can now compute the function $g_n(t)$ of equation (6.5). Set

$$g_1(t) = |a_1| g(a_1 t - a_1 b - 1),$$

$$g_2(t) = |a_2| g(a_2 t - 1) * g_1(t),$$

$$\cdot \quad \cdot \quad \cdot \quad \cdot \quad \cdot \quad \cdot \quad \cdot \quad \cdot \quad \cdot$$

$$g_n(t) = |a_n| g(a_n t - 1) * g_{n-1}(t).$$

Then equation (6.5) holds with this function $g_n(t)$, and since $g(t) \geq 0$ it is clear that $g_n(t) \geq 0$. The integral converges absolutely for $|\sigma| < |a_1|$.

Inverting the integral (6.5) gives

$$(6.6) \qquad g_n(t) = \frac{1}{2\pi i} \int_{-i\infty}^{i\infty} \frac{e^{st}}{P_n(s)} \, ds, \qquad -\infty < t < \infty.$$

This integral converges absolutely when $n \geq 2$ since

$$(6.7) \qquad |P_n(iy)|^2 = \prod_{k=1}^{n} \left(1 + \frac{y^2}{a_k^2}\right) \geq \left(1 + \frac{y^2}{a_1^2}\right)\left(1 + \frac{y^2}{a_2^2}\right).$$

Inequality (6.7) also enables us to apply Lebesgue's limit theorem to take the limit under the integral sign in (6.6), since

$$\int_{-\infty}^{\infty} \frac{e^{iyt}}{P_n(iy)} \, dy \ll \int_{-\infty}^{\infty} \frac{dy}{\left(1 + \dfrac{y^2}{a_1^2}\right)^{1/2}\left(1 + \dfrac{y^2}{a_2^2}\right)^{1/2}} < \infty, \quad n \geq 2.$$

Thus

$$\lim_{n \to \infty} g_n(t) = \frac{1}{2\pi i} \int_{-i\infty}^{i\infty} \frac{e^{st}}{E(s)} \, ds = G(t),$$

and $G(t) \geq 0$, as we wished to prove. The basic fact used in the latter part of this proof is that the convolution of two frequency functions is again a frequency function.

7. Inversion

We are now in a position to invert any convolution transform whose kernel has a bilateral Laplace transform whose reciprocal is in E_0. We define the inversion operator $E(D)$ in the following natural way.

Definition 7. If $E(s) \in E_0$,

$$E(s) = e^{bs} \prod_{k=1}^{\infty} \left(1 - \frac{s}{a_k}\right) \exp\left(\frac{s}{a_k}\right),$$

then

$$E(D)f(x) = \lim_{n \to \infty} \prod_{k=1}^{n} \left(1 - \frac{D}{a_k}\right) f\left(x + b + \sum_{k=1}^{n} \frac{1}{a_k}\right).$$

The operator $E(D)$ may be described accordingly as a linear differential operator with constant coefficients, of infinite order, plus certain translations. In particular the translations can be accumulated into a single one through the amount $b + \sum_{1}^{\infty} a_k^{-1}$ in the special case for which that series converges.

Theorem 7.1.

1. $E(s) \in E_0$, $E(s) = e^{bs} \prod_{k=1}^{\infty} \left(1 - \frac{s}{a_k}\right) \exp\left(\frac{s}{a_k}\right)$;

2. $G(t) = \dfrac{1}{2\pi i} \displaystyle\int_{-i\infty}^{i\infty} \frac{e^{st}}{E(s)} \, ds, \qquad -\infty < t < \infty$;

3. $\phi(t) \in \text{B.C}, \qquad -\infty < t < \infty$;

4. $f(x) = G * \phi = \displaystyle\int_{-\infty}^{\infty} G(x - t)\phi(t) \, dt$

$\Rightarrow \qquad\qquad E(D)f(x) = \phi(x), \qquad -\infty < x < \infty.$

As in §6 we define

$$P_n(s) = e^{bs} \prod_{k=1}^{n} \left(1 - \frac{s}{a_k}\right) \exp\left(\frac{s}{a_k}\right), \qquad E_n(s) = \prod_{k=n+1}^{\infty} \left(1 - \frac{s}{a_k}\right) \exp\left(\frac{s}{a_k}\right),$$

so that $E(s) = P_n(s)E_n(s)$. We begin by applying the operator $P_n(D)$ to the kernel $G(x)$. Set

$$(7.1) \qquad G_n(t) = \frac{1}{2\pi i} \int_{-i\infty}^{i\infty} \frac{e^{st}}{E_n(s)} \, ds, \qquad n = 1, 2, \ldots .$$

The functions $G_n(t)$ are Laguerre–Pólya kernels and have all the properties described in Theorem 6.4 since the functions $E_n(s)$ clearly are members of E_0. The integrals (7.1) converge absolutely for $-\infty < t < \infty$ by Theorem 6.2. Since formal differentiation gives

$$D^k G(t) = \frac{1}{2\pi i} \int_{-i\infty}^{i\infty} \frac{e^{st} s^k}{E(s)} \, ds$$

and since

$$e^{aD} G(t) = G(t + a) = \frac{1}{2\pi i} \int_{-i\infty}^{i\infty} \frac{e^{st} e^{as}}{E(s)} \, ds,$$

we have the formal result

$$(7.2) \quad P_n(D) G(t) = \frac{1}{2\pi i} \int_{-i\infty}^{i\infty} \frac{e^{st} P_n(s)}{E(s)} \, ds = \frac{1}{2\pi i} \int_{-i\infty}^{i\infty} \frac{e^{st}}{E_n(s)} \, ds,$$

$$P_n(D) G(t) = G_n(t).$$

To justify equation (7.2) we have only to apply the classical theorem on differentiation under an integral sign by showing that the resulting integral (7.2) converges uniformly on every compact set of the t axis. Actually it does so over the entire axis since

$$\int_{-\infty}^{\infty} \frac{e^{iyt}}{E_n(iy)} \, dy \ll \int_{-\infty}^{\infty} \frac{dy}{|E_n(iy)|} < \infty, \qquad -\infty < t < \infty.$$

The dominant integral converges by Theorem 6.2.

We next apply the operator $P_n(D)$ to $f(x)$:

$$(7.3) \quad P_n(D) f(x) = P_n(D) \int_{-\infty}^{\infty} G(x - t)\phi(t) \, dt = \int_{-\infty}^{\infty} G_n(x - t)\phi(t) \, dt$$

provided again that the resulting integral is uniformly convergent on compact sets of the x axis, say on $|x| \leq A$ for arbitrary A. If $|\phi(t)| < M, -\infty < t < \infty$, then

$$\left| \int_{-\infty}^{-R} G_n(x - t)\phi(t) \, dt \right| \leq M \int_{x+R}^{\infty} G_n(t) \, dt \leq M \int_{-A+R}^{\infty} G_n(t) \, dt,$$

$$|x| \leq A.$$

The dominant integral is independent of x and is small for large R, so that the integral

$$\int_{-\infty}^{0} G_n(x - t)\phi(t) \, dt$$

does converge uniformly on $|x| \leq A$. A similar proof applies to the integral extended over $(0, \infty)$. Thus equation (7.3) is valid.

Finally we must show that $P_n(D)f(x) \to \phi(x)$ as $n \to \infty$, or that

$$(7.4) \quad \lim_{n \to \infty} \int_{-\infty}^{\infty} G_n(x - t)[\phi(t) - \phi(x)]\, dt$$

$$= \lim_{n \to \infty} \int_{-\infty}^{\infty} G_n(t)[\phi(x - t) - \phi(x)]\, dt = 0.$$

Here we have used the fact that $G_n(x)$ is itself a frequency function and has all the properties in the conclusion of Theorem 6.4.

For an arbitrary $\delta > 0$ we break the integral (7.4) into two parts corresponding to the ranges of integration $|t| < \delta$ and $|t| \geq \delta$, denoting the parts by I_n and J_n, respectively. Again by Theorem 6.4 we have for a fixed x

$$(7.5) \quad |I_n| \leq \max_{|t| \leq \delta} |\phi(x - t) - \phi(x)| \int_{-\delta}^{\delta} G_n(t)\, dt$$

$$\leq \max_{|t| \leq \delta} |\phi(x - t) - \phi(x)|.$$

Since $|t| \geq \delta \Rightarrow 1 \leq t^2/\delta^2$ we have

$$|J_n| \leq 2M \int_{|t| \geq \delta} G_n(t)\, dt \leq \frac{2M}{\delta^2} \int_{|t| \geq \delta} t^2 G_n(t)\, dt$$

$$(7.6) \qquad \leq \frac{2M}{\delta^2} \int_{-\infty}^{\infty} t^2 G_n(t)\, dt = \frac{2M}{\delta^2} \sum_{k=n}^{\infty} \frac{1}{a_k^2}.$$

Here we have used the fact that the mean of $G_n(t)$ is zero and have computed its variance by Theorem 6.4. From (7.5) and (7.6) it follows that

$$(7.7) \qquad \overline{\lim_{n \to \infty}} \, |I_n + J_n| \leq \max_{|t| \leq \delta} |\phi(x - t) - \phi(x)|.$$

Since this limit superior is independent of δ and since the right-hand side of (7.7) $\to 0$ as $\delta \to 0$, the left side must also be zero and equation (7.4) is proved.

It is apparent from the proof that the continuity of $\phi(x)$ is used only at the point x where the inversion of the transform is to be accomplished. In fact, the theorem can be improved drastically by removing all local conditions on $\phi(x)$ and requiring only that the integral $G * \phi$ should converge. The conclusion $E(D)f(x) = \phi(x)$ is still true for almost all x. The situation is in marked contrast to that involved in Theorem 7.3, Chapter 5, for example, where some strong local condition like bounded variation is required. It is the positiveness of the kernels $G_n(t)$ that makes the difference. See I. I. Hirschman and D. V. Widder [1955, p. 139].

In one of our applications we shall need a slightly more general result than that contained in Theorem 7.1. If $E(D)$ is replaced by $\overline{\lim}_{n\to\infty} \exp(\varepsilon_n D)P_n(D)$, where $\varepsilon_n \to 0$ as $n \to \infty$, the new operator still acomplishes the inversion.

Theorem 7.2.

$$1. \quad \varepsilon_n = o(1), \qquad n \to \infty;$$

$$2. \quad \text{All conditions of Theorem 7.1}$$

$$\Rightarrow \qquad \lim_{n \to \infty} \exp(\varepsilon_n D)\, P_n(D)f(x) = \phi(x), \qquad -\infty < x < \infty.$$

Since $\exp(\varepsilon_n D)$ means a translation through ε_n we now must prove that

$$(7.8) \qquad \lim_{n \to \infty} \int_{-\infty}^{\infty} G_n(t)\phi(x - t + \varepsilon_n)\, dt = \phi(x).$$

As before, we break the integral (7.8) into two parts corresponding to $|t| < \delta$ and $|t| \geq \delta$. Inequality (7.5) now becomes

$$|I_n| \leq \max_{|t| \leq \delta} |\phi(x - t + \varepsilon_n) - \phi(x)|.$$

By the continuity of $\phi(t)$ we have

$$\overline{\lim_{n \to \infty}} \; |I_n| \leq \max_{|t| \leq \delta} |\phi(x - t) - \phi(x)|,$$

and the rest of the proof follows as before.

8. The Laplace Transform as a Convolution

We now apply the general inversion formula of §7 to the special case of Example A of §2, thus recapturing the inversion of the Laplace transform given in Chapter 6. We need a preliminary result involving linear differential operators.

Theorem 8.1.

$$1. \quad F(x) \in C^n, \qquad 0 < x < \infty$$

$$\Rightarrow \qquad A. \quad \prod_{k=1}^{n} \left(1 - \frac{D}{k}\right) e^x F(e^x) = \frac{(-1)^n}{n!} \, e^{(n+1)x} F^{(n)}(e^x);$$

$$B. \quad \prod_{k=1}^{n} \left(1 + \frac{D}{k}\right) e^{-nx} F(e^x) = \frac{1}{n!} \, F^{(n)}(e^x).$$

To prove A note first that for $k = 1, 2, 3, \ldots$,

$$\left(1 - \frac{D}{k}\right) e^{kx} F^{(k-1)}(e^x) = -\frac{1}{k} \, e^{(k+1)x} F^{(k)}(e^x).$$

Hence if we operate successively, in the order $k = 1, 2, 3, \ldots$ we obtain the result at once. Since the coefficients of the differential operator are constants the order of applying the factors is not involved, is only important for quick computation. For B, simple calculation gives for $k = n, n - 1, \ldots, 1$,

$$\left(1 + \frac{D}{k}\right) e^{-kx} F(e^x) = \frac{1}{k} \, e^{-(k-1)x} F'(e^x).$$

This time it is convenient to apply the separate factors of B in the opposite order, $k = n, n - 1, \ldots, 1$. Conclusion B results at once.

We now show that the Laplace transform, Example A of §2, may be inverted as a special case of Theorem 7.2. As noted in §2, the transform (2.2) is equivalent to (2.1) with kernel

$$G(x) = e^x \exp(-e^x).$$

We show first that this is a Laguerre–Pólya kernel:

$$\int_{-\infty}^{\infty} e^{-st} e^t \exp(-e^t) \, dt = \int_{0}^{\infty} e^{-t} t^{-s} \, dt = \Gamma(1 - s).$$

The familiar product expansion of $\Gamma(s)$ is

$$\frac{1}{\Gamma(s)} = e^{\gamma s} s \prod_{k=1}^{\infty} \left(1 + \frac{s}{k}\right) e^{-s/k}.$$

From the identity

$$\frac{1}{\Gamma(s)\Gamma(1-s)} = \frac{\sin \pi s}{\pi} = s \prod_{k=1}^{\infty} \left(1 - \frac{s^2}{k^2}\right)$$

we obtain

$$\frac{1}{\Gamma(1-s)} = e^{-\gamma s} \prod_{k=1}^{\infty} \left(1 - \frac{s}{k}\right) e^{s/k}.$$

Here γ is Euler's constant,

$$(8.1) \qquad \gamma = \lim_{n \to \infty} \left[\sum_{k=1}^{n} \frac{1}{k} - \log n\right].$$

Clearly $1/\Gamma(1-s) \in E_0$. The constants of Definition 4.2 are $b = -\gamma$, $a_k = k$,

$$\sum_{k=1}^{\infty} \frac{1}{k^2} < \infty.$$

We apply Theorem 7.2 with

$$\varepsilon_n = \gamma + \log n - \sum_{k=1}^{n} \frac{1}{k},$$

$$P_n(D) = e^{-\gamma D} \prod_{k=1}^{n} \left(1 - \frac{D}{k}\right) e^{D/k}.$$

Then $\varepsilon_n \to 0$ as $n \to \infty$ by (8.1) and

$$\exp(\varepsilon_n D)P_n(D) = n^D \prod_{k=1}^{n} \left(1 - \frac{D}{k}\right).$$

Of course

$$n^D f(e^x) = e^{D \log n} f(e^x) = f(ne^x).$$

Using the notation of Example A of §2, Theorem 7.2 gives

$$\lim_{n \to \infty} n^D \prod_{k=1}^{n} \left(1 - \frac{D}{k}\right) e^x F(e^x) = \Phi(e^{-x}).$$

By Theorem 8.1, conclusion A, we have

$$(8.2) \quad \lim_{n \to \infty} \exp(\varepsilon_n D) P_n(D) e^x F(e^x) = \lim_{n \to \infty} (-1)^n \frac{n^{n+1}}{n!} e^{(n+1)x} F^{(n)}(ne^x)$$

$$= \Phi(e^{-x}).$$

Setting $e^{-x} = y$ equation (8.2) becomes

$$\lim_{n \to \infty} \frac{(-1)^n}{n!} \left(\frac{n}{y}\right)^{n+1} F^{(n)}\left(\frac{n}{y}\right) = \Phi(y), \qquad 0 < y < \infty.$$

We have thus proved

Theorem 8.2.

$$1. \quad \Phi(t) \in \text{B.C}, \qquad 0 < t < \infty;$$

$$2. \quad F(x) = \int_0^\infty e^{-xt} \Phi(t) \, dt$$

$$\Rightarrow \quad \lim_{n \to \infty} (-1)^n \left(\frac{n}{t}\right)^{n+1} F^{(n)}\left(\frac{n}{t}\right) = \Phi(t), \qquad 0 < t < \infty.$$

Except for an unimportant difference in hypothesis this is Theorem 3 of Chapter 6.

9. The Stieltjes Transform as a Convolution

We turn next to Example B of §2. If we assume only that $\Phi(t) \in \text{B.C}$, as in Theorem 7.1, the Stieltjes transform need not converge. For example if $\Phi(t) \equiv 1$, $F(x)$ is undefined. However, to apply Theorem 7.1 we have only to make a preliminary differentiation. Thus we have

$$(9.1) \qquad -F'(x) = \int_0^\infty \frac{\Phi(t)}{(x + t)^2} \, dt,$$

and this assumes the form (2.1) if we set

$$f(x) = -e^x F'(e^x), \qquad G(x) = e^x (e^x + 1)^{-2}, \qquad \phi(t) = \Phi(e^t).$$

For the inversion function $E(s)$ we have the equation

$$\frac{1}{E(s)} = \int_{-\infty}^\infty \frac{e^{-st} e^t}{(e^t + 1)^2} \, dt = \int_0^\infty \frac{x^{-s}}{(x + 1)^2} \, dx = \frac{\pi s}{\sin \pi s}.$$

Hence the symbolic statement of Theorem 7.1 becomes

$$\frac{-\sin \pi D}{\pi D} e^x F'(e^x) = \Phi(e^x).$$

This symbolic statement, interpreted as usual by the infinite product expansion of the inversion function, becomes

$$(9.2) \qquad -\lim_{n \to \infty} \prod_{k=1}^{n} \left(1 + \frac{D}{k}\right) \prod_{j=1}^{n} \left(1 - \frac{D}{j}\right) e^x F'(e^x) = \Phi(e^x).$$

But by conclusion A of Theorem 8.1

$$(9.3) \qquad \prod_{k=1}^{n} \left(1 - \frac{D}{k}\right) e^x F'(e^x) = \frac{(-1)^n}{n!} e^{(n+1)x} F^{(n+1)}(e^x).$$

In order to apply conclusion B of Theorem 8.1 we temporarily introduce a new function H by the equation

$$(9.4) \qquad e^{-nx} H(e^x) = \frac{(1)^n}{n!} e^{(n+1)x} F^{(n+1)}(e^x).$$

Then

$$\prod_{k=1}^{n} \left(1 + \frac{D}{k}\right) e^{-nx} H(e^x) = \frac{1}{n!} H^{(n)}(e^x).$$

Now replace e^x by a new variable y and eliminate H between equations (9.3) and (9.4). We obtain

$$-\prod_{k=1}^{n} \left(1 + \frac{D}{k}\right) \prod_{j=1}^{n} \left(1 - \frac{D}{j}\right) e^x F'(e^x) = \frac{(-1)^{n+1}}{n!\,n!} \cdot D_y^n y^{2n+1} D_y^{n+1} F(y),$$

where D_y stands for differentiation with respect to y. In the notation of Definition 4 of Chapter 6 equation (9.2) becomes

$$\lim_{n \to \infty} M_{n,y}[F(y)] = \Phi(y),$$

and we are led to the following theorem.

Theorem 9.1.

> 1. $\Phi(t) \in$ B.C, $\qquad 0 < t < \infty$;
>
> 2. $F(x) = \displaystyle\int_0^{\infty} \frac{\Phi(t)}{x+t}\,dt \quad$ converges, $\qquad 0 < x < \infty$

$$(9.5) \quad \Rightarrow \quad \lim_{n \to \infty} \frac{(-1)^{n+1}}{n!\,n!} [x^{2n+1} F^{(n+1)}(x)]^{(n)} = \Phi(x), \qquad 0 < x < \infty.$$

Note that we have now postulated the convergence of the given Stieltjes transform. As noted above, hypothesis 1 alone does not produce it. The only part of the theorem left unproved is that equation (9.1) follows from the hypotheses 1 and 2. But for $0 < \delta \leq x < \infty$

$$\int_0^\infty \frac{\Phi(t)}{(x+t)^2}\, dt \ll \operatorname*{u.b.}_{0<t<\infty} |\Phi(t)| \int_0^\infty \frac{dt}{(\delta+t)^2} < \infty,$$

so that the integral (9.1) converges uniformly in $\delta \leq x < \infty$, and consequently differentiation under the integral sign (9.5) is permissible.

If in Example D of §2 we replace x by $\sqrt{x}$ and t by $\sqrt{t}$ we obtain

$$F(\sqrt{x}) = \frac{1}{\pi} \int_0^\infty \frac{\Phi(\sqrt{t})}{x+t}\, dt,$$

so that the potential transform is another form of the Stieltjes transform. We are thus led to the following result.

Theorem 9.2.

$$1. \quad \Phi(t) \in \text{B.C}, \qquad 0 < t < \infty;$$

$$2. \quad F(x) = \frac{2}{\pi} \int_0^\infty \frac{t\Phi(t)}{x^2+t^2}\, dt \quad \text{converges,} \qquad 0 < x < \infty,$$

$$(9.6) \quad \Rightarrow \quad \lim_{n \to \infty} \frac{(-1)^{n+1}}{n!\,n!} D^n[x^{2n+1}D^{n+1}F(\sqrt{x})] = \frac{\Phi(\sqrt{x})}{\pi},$$

$$0 < x < \infty.$$

An example is provided by the pair $\Phi = 1/t$, $F = 1/x$. In this case the differentiations may be completed explicitly, and equation (9.6) becomes

$$\lim_{n \to \infty} \frac{(2n+1)!(2n)!}{(n!)^4 2^{4n+1}} \frac{1}{\sqrt{x}} = \frac{1}{\pi\sqrt{x}}, \qquad 0 < x < \infty,$$

and this may be verified by use of Stirling's formula. In general the calculations involved in formula (9.6) are not feasible and other methods of inversion are more useful. See Chapter 9 and Theorem 14.2 of Chapter 5.

10. Summary

The inversion function $E(s)$ for the convolution transform

$$(10.1) \qquad f(x) = \int_{-\infty}^{\infty} G(x - t)\phi(t)\, dt$$

is the reciprocal of the bilateral Laplace transform of the kernel G,

$$(10.2) \qquad \frac{1}{E(s)} = \int_{-\infty}^{\infty} e^{-st}G(t)\, dt.$$

That is,

$$(10.3) \qquad E(D)f(x) = \phi(x),$$

where D stands for differentiation with respect to x.

The symbolic equation (10.3) can be realized practically for a large class of kernels G. These derive by equation (10.2) from functions $E(s)$ which are in the Laguerre–Pólya class, that is, those functions which are uniform limits of polynomials with real roots only. We have here treated a subclass, those functions $E(s)$ which have an expansion

$$E(s) = e^{bs} \prod_{k=1}^{\infty} \left(1 - \frac{s}{a_k}\right)\exp\left(\frac{s}{a_k}\right), \qquad b,\ a_k \text{ real},$$

$$\sum_{k=1}^{\infty} 1/a_k^2 < \infty.$$

Then equation (10.3) is interpreted to mean

$$(10.4) \qquad \lim_{n\to\infty} e^{bD} \prod_{k=1}^{n} \left(1 - \frac{D}{a_k}\right)\exp\left(\frac{D}{a_k}\right) f(x) = \phi(x),$$

where $e^{aD}f(x) = f(x + a)$.

The Laplace transform has inversion function $E(s) = 1/\Gamma(1 - s)$ and the Stieltjes transform, $E(s) = (\sin \pi s)/(\pi s)$. In each case equation (10.4) recaptures the inversion of Chapter 6.

EXERCISES

1. If $g(x) = e^x (x < 0), = O\,(x \geq 0)$, show that $g * g = |x|e^x\ (x < 0)$, $= O(x \geq 0)$. If $\phi \in$ B.C on $(-\infty, \infty)$ show that
    $$(1 - D)^2\, g * g * \phi = \phi, \qquad -\infty < x < \infty.$$

2. Generalize the previous exercise.

3. If $h(x) = e^{-|x|}/2$ and $\phi \in$ B.C on $(-\infty, \infty)$ show that

$$(1 - D^2) h * \phi = \phi, \qquad -\infty < x < \infty.$$

4. If $h(x)$ is the function of Exercise 3, compute $h * h$ and prove that

$$(1 - D^2)^2 h * h * \phi = \phi, \qquad -\infty < x < \infty.$$

5. Using Definition 3.2 compute the inversion functions for the kernels of Exercises 1 and 2, designating the regions of convergence of the Laplace integrals involved.

6. Same problems for Exercises 3 and 4.

7. If $g(x)$ and $h(x)$ are the functions of Exercises 1 and 3 compute $g * h$ and the corresponding inversion function.

$$\text{Ans.} \quad e^{-|x|}(|x| - x + 1)/4.$$

8. By differentiating equation (4.5) (or otherwise) show that

$$\int_{-\infty}^{\infty} y^2 k(y, t)\, dy = 2t, \qquad \int_{-\infty}^{\infty} y^4 k(y, t)\, dy = 12t^2, \qquad t > 0,$$

$$k(x, t) * x^4 = \int_{-\infty}^{\infty} k(x - y, t) y^4\, dy = x^4 + 12x^2 t + 12t^2, \qquad t > 0.$$

9. If for any real t we define

$$(11.1) \qquad \exp(tD^2) f(x) = \sum_{k=0}^{\infty} \frac{t^k}{k!} f^{(2k)}(x),$$

show that

$$\exp(tD^2) x^4 = k(x, t) * x^4, \qquad t > 0.$$

10. Using Definition 3.2 show that the inversion function for the kernel $k(x, t_0)$, $t_0 > 0$, is $\exp(-t_0 s^2)$. Using (11.1) prove

$$\exp(-tD^2)(k(x, t) * x^4) = x^4,$$

as the symbolic equation (3.1) would predict.

11. Generalize exercises 8, 9, 10:

$$\int_{-\infty}^{\infty} k(y, t)y^{2n}\, dy = \frac{(2n)!\, t^n}{n!},$$

$$k(x, t) * x^n = n! \sum_{k=0}^{[n/2]} \frac{t^k}{k!} \frac{x^{n-2k}}{(n-2k)!},$$

$$\exp(-tD^2)(k(x, t) * x^n) = x^n.$$

12. Using (10.1) show that for any polynomial $P(x)$

$$\exp(tD^2)P(x) = k(x, t) * P(x), \qquad t > 0.$$

[Hint: Expand $P(x - y)$ in Maclaurin's series.]

13. If $\phi(t) \in$ B.C on $(0, \infty)$, p is a positive integer, and

$$f(x) = (p + 1)! \int_0^{\infty} \frac{\phi(t)\, dt}{(x + t)^{p+2}},$$

show that

$$\lim_{k \to \infty} \frac{(-1)^{k-p}}{k!\, k!} [t^{2k+1} f^{(k-p)}(t)]^{(k)} = \phi(t), \qquad 0 < t < \infty.$$

Check for the case $\phi(t) \equiv 1$.

8 *Tauberian Theorems*

1. Introduction

We recall that a series

$$(1.1) \qquad \sum_{k=0}^{\infty} a_k$$

is *summable* (A), in the sense of Abel, to the number A if the power series

$$f(x) = \sum_{k=0}^{\infty} a_k x^k$$

converges for $|x| < 1$ and if $f(1-) = A$. The classical theorem of Abel is a *regularity* or *consistency* result to the effect that the convergence of (1.1) to A implies its Abel summability to the same value. The example $a_k = (-1)^k$ shows that the converse statement is not true. It is this very fact that makes summability of divergent series a useful concept.

A "corrected" converse is, however, true. For example, summability (A) implies convergence if one of the following conditions holds:

$$\text{A.} \quad a_k \geq 0,$$

$$\text{B.} \quad a_k = o\left(\frac{1}{k}\right), \qquad k \to \infty,$$

$$\text{C.} \quad a_k = O\left(\frac{1}{k}\right), \qquad k \to \infty.$$

The first of these results is trivial as a consequence of the following inequalities:

$$\sum_{k=0}^{n} a_k x^k \leq f(x), \qquad 0 < x < 1, \quad n = 0, 1, 2, \ldots,$$

$$(1.2) \qquad s_n = \sum_{k=0}^{n} a_k \leq \lim_{x \to 1-} f(x).$$

Since the partial sums $s_n \in \uparrow$, they must approach a limit if $f(1-)$ is finite; hence summability implies convergence. The case B was first proved by A. Tauber [1897], case C by J. E. Littlewood [1910]. The latter result and closely related ones might be said to form the underlying theme of the present chapter.

We recall further that series (1.1) is *summable* (C), in the sense of Cesàro, to the number A if

$$(1.3) \qquad \lim_{n \to \infty} \frac{s_0 + s_1 + \cdots + s_n}{n + 1} = A.$$

This method of summability is also regular. (See Exercise 1 in this chapter.) The same example used above shows that summability (C) does not imply convergence. Again a "corrected" converse can be proved. For example, G. H. Hardy [1910, p. 301] proved that condition C above is a suitable correction. We shall prove this later.

Note that

$$f(x) = \frac{s_0 + s_1 x + s_2 x^2 + \cdots}{1 + x + x^2 + \cdots},$$

so that $f(x)$ may be regarded as a weighted average, with weights x^k, of all s_k. Abel summability results in the limiting case when all weights are unity. Similarly, equation (1.3) asserts, in a sense, that the arithmetic average of "all" s_k is A. These two examples are typical of summability theory. In general we say that a theorem is *Abelian* if it asserts something about an average of a sequence from a hypothesis about its ordinary limit; it is *Tauberian* if conversely the implication goes from average to limit. Usually the latter results must have in the hypothesis some supplementary condition, like A, B, or C above. In this chapter we shall be dealing chiefly with Tauberian theorems involving integrals, which are the continuous analogs of series. After such results are proved it is then a fairly simple matter to obtain, as special cases, the classical ones about series. See §7.

We conclude this section with the proof of a typical Abelian result, which will be useful later.

Theorem 2.1.

$$1. \quad f(x) = \int_0^\infty e^{-xt} a(t)\, dt, \qquad 0 < x < \infty;$$

$$2. \quad a(t) \sim At^\gamma, \qquad t \to \infty, \quad \gamma > -1$$

$$\Rightarrow \qquad f(x) \sim \frac{A\Gamma(\gamma + 1)}{x^{\gamma + 1}}, \qquad x \to 0+.$$

Throughout this chapter we shall assume without further statement that $a(t) \in L$ in $(0, R)$ for every $R > 0$. Since

$$\int_0^\infty e^{-xt} t^\gamma\, dt = \frac{\Gamma(\gamma + 1)}{x^{\gamma + 1}}, \qquad x > 0,$$

it is no restriction to assume that $A = 0$, $a(t) = o(t^\gamma)$. Then to an arbitrary $\varepsilon > 0$ there corresponds an R such that

$$|a(t)| < \varepsilon t^\gamma, \qquad R \leq t < \infty.$$

Hence

$$x^{\gamma + 1}|f(x)| < x^{\gamma + 1} \int_0^R e^{-xt}|a(t)|\, dt + x^{\gamma + 1}\varepsilon \int_R^\infty e^{-xt} t^\gamma\, dt$$

$$< x^{\gamma + 1} \int_0^R |a(t)|\, dt + \varepsilon\Gamma(\gamma + 1);$$

$$\varlimsup_{x \to 0+} x^{\gamma + 1}|f(x)| \leq \varepsilon\Gamma(\gamma + 1).$$

It follows that

$$f(x) = o(x^{-\gamma-1}), \qquad x \to 0+,$$

as we wished to prove.

2. Integral Analogs

In this chapter we shall be primarily concerned with Tauberian theorems involving Laplace integrals, and these may be regarded as continuous analogs of Tauber's original series theorem, or generalizations thereof. We say that the integral

$$(2.1) \qquad \int_0^\infty a(t)\, dt$$

is *summable* (A) to the value A if the Laplace integral

$$(2.2) \qquad f(x) = \int_0^\infty e^{-xt} a(t)\, dt$$

converges for $x > 0$ and if $f(1-) = A$. It is *summable* (C) if

$$(2.3) \qquad A(x) = \int_0^x a(t)\, dt, \qquad \lim_{x \to \infty} \frac{1}{x} \int_0^x A(t)\, dt = A.$$

Our first result is the regularity theorem.

Theorem 2.1.

$$1. \quad f(x) = \int_0^\infty e^{-xt} a(t)\, dt, \qquad 0 < x < \infty;$$

$$2. \quad \int_0^\infty a(t)\, dt = A$$

$$\Rightarrow \qquad f(0+) = A.$$

That is, the convergence of (2.1) implies its Abel summability to the same value. This result is an immediate consequence of Theorem 4.1 of Chapter 5.

A converse of this, corresponding to Condition A of §1 follows.

Theorem 2.2.

$$1. \quad f(x) = \int_0^\infty e^{-xt} a(t)\, dt, \qquad 0 < x < \infty;$$

$$2. \quad f(0+) = A;$$

$$3. \quad a(t) \geq 0, \qquad 0 < t < \infty$$

$$\Rightarrow \qquad \int_0^\infty a(t)\, dt = A.$$

Analogous to the corresponding proof for series given in §1, this result follows easily from the inequality

$$A(R) = \lim_{x \to 0+} \int_0^R e^{-xt} a(t)\, dt \leq \varliminf_{x \to 0+} f(x).$$

The example $a(t) = \cos t$ shows that hypothesis 3 cannot be omitted. For, in this case $f(x) = x/(x^2 + 1)$, $f(0+) = 0$, but the integral (2.1) diverges.

Our next result is considered Abelian though it is a comparison of two kinds of averages. It states that Cesàro summability implies Abel summability.

Theorem 2.3.

$$1. \quad f(x) = \int_0^\infty e^{-xt} a(t)\, dt, \qquad 0 < x < \infty;$$

$$(2.4) \qquad 2. \quad \int_0^x a(t)\, dt \sim Ax, \qquad x \to \infty$$

$$(2.5) \quad \Rightarrow \qquad f(x) \sim \frac{A}{x}, \qquad x \to 0+.$$

If $a(t)$ is an integral

$$a(t) = \int_0^t b(y)\, dy,$$

then it is the integral

$$\int_0^\infty b(t)\, dt$$

that is summable (C) by hypothesis 2 and which is summable (A) by (2.5). However, the theorem as stated, in terms of $a(t)$ rather than $b(t)$, suggests a different interpretation. The integral (2.1) diverges to $+\infty$ ($A > 0$), but the partial integrals $A(x)$ become infinite in the prescribed way (2.4). The conclusion is that $f(x)$ also becomes infinite, but at a resultant rate (2.5), as $x \to 0+$.

This result is essentially contained in Theorem 1. For, an integration by parts gives

$$f(x) = x \int_0^\infty e^{-xt} A(t)\, dt, \qquad x > 0,$$

and (2.4) becomes $A(x) \sim Ax$, $x \to \infty$. The conclusion of Theorem 1, $\gamma = 1$, applied to the integral $f(x)/x$ now yields the desired result.

An example to show that the theorem is not reversible is

$$(2.6) \qquad a(t) = 1 + \sin t + t \cos t, \qquad A(t) = t + t \sin t$$

$$f(x) = \frac{1}{x} + \frac{2x^2}{(x^2 + 1)^2}.$$

Here $f(x) \sim 1/x$, $x \to 0+$, but $A(t)/t$ approaches no limit as $t \to \infty$. Accordingly, to obtain a corrected converse we must expect to impose some additional condition on $a(t)$. It turns out that boundedness is a suitable one. Note that $a(t)$ as defined by (2.6) is not bounded.

Theorem 2.4.

$$1. \quad f(x) = \int_0^\infty e^{-xt} a(t)\, dt, \qquad 0 < x < \infty;$$

$$2. \quad a(t) \in B \quad (\text{bounded}), \qquad 0 < t < \infty;$$

$$3. \quad f(x) \sim \frac{A}{x}, \qquad x \to 0+$$

$$(2.7) \quad \Rightarrow \qquad \int_0^x a(t)\, dt \sim Ax, \qquad x \to \infty.$$

We shall obtain this result in the next section as a special case of a much more general theorem. But let us illustrate it here by the example

$$a(t) = \int_0^t \frac{\sin y}{y}\, dy \sim \frac{\pi}{2}, \qquad t \to +\infty.$$

The Laplace transform of $a(t)$ is (Exercise 14)

$$(2.8) \qquad f(x) = \frac{1}{x} \tan^{-1} \frac{1}{x} \sim \frac{\pi}{2x}, \qquad x \to 0+.$$

Since $a(t) \in B$ we may apply the theorem to obtain (2.7) with $A = \pi/2$. But this may be verified directly since

$$\int_0^x a(t)\, dt = \int_0^x (x - y) \frac{\sin y}{y}\, dy = x \int_0^x \frac{\sin y}{y}\, dy + \cos x - 1,$$

$$\lim_{x \to \infty} \frac{1}{x} \int_0^x a(t)\, dt = \int_0^\infty \frac{\sin y}{y}\, dy = \frac{\pi}{2}.$$

Exercise 3 shows that hypothesis 2 is essential here.

3. A Basic Tauberian Theorem

We observe that hypothesis 3 of Theorem 2.4 may be written as

$$(3.1) \qquad \frac{1}{x} \int_0^\infty g\!\left(\frac{t}{x}\right) a(t)\, dt \sim A \int_0^\infty g(t)\, dt, \qquad x \to \infty$$

if $g(t) = e^{-t}$ and that the conclusion (2.7) takes the same form if $g(t)$ is replaced by $h(t) = 1$ on $(0, 1)$, $h(t) = 0$ on $(1, \infty)$. Thus Theorem 2.4 states that, under the restriction of boundedness on $a(t)$, if (3.1) holds for one "kernel" $g(x)$ it also holds for at least one other kernel $h(t)$. Our basic theorem will show that the conclusion will hold for *many* kernels $h(t)$ even if $g(t)$ is not e^{-t}. The essential property of e^{-t} for our purpose is contained in the familiar uniqueness property of the Laplace transform: the vanishing of the generating function implies that of the determining function. We dignify this property in a kernel $g(t)$ by a notation as follows.

Definition 3.1. A function $g(x) \in U$ (has the uniqueness property) on $(0, \infty)$ if $g(x) \in L$ on $(0, \infty)$ and if the equation

$$\int_0^\infty g\!\left(\frac{t}{x}\right) a(t)\, dt \equiv 0, \qquad 0 < x < \infty$$

for $a(t) \in B.C$ implies $a(t) \equiv 0$, $0 < t < \infty$.

Clearly an exponential change of variables converts the integral (3.1) into a convolution, the range of integration becoming the whole real axis. Our uniqueness property takes the following form for the interval $(-\infty, \infty)$.

Definition 3.2. A function $g(x) \in U$ on $(-\infty, \infty)$ if $g(x) \in L$ on $(-\infty, \infty)$ and if the equation

$$\int_{-\infty}^{\infty} g(x-t)\, a(t)\, dt \equiv 0, \qquad 0 < x < \infty$$

for $a(t) \in \mathrm{B.C}$ implies $a(t) \equiv 0$, $-\infty < t < \infty$.

Note that if $g(x) \in U$ on $(0, \infty)$ then $g(e^x)e^x \in U$ on $(-\infty, \infty)$. For example $e^{-x} \in U$ on $(0, \infty)$ and $\exp(-e^x)e^x \in U$ on $(-\infty, \infty)$. On the other hand a function $g(x)$ which is constant over a finite interval and is zero elsewhere $\notin U$ on $(-\infty, \infty)$. This can be seen by choosing $a(t)$ as a suitable periodic function. (See Exercise 12 of this chapter.)

We now state our basic Tauberian Theorem as follows.

Theorem 3.1.

$$1. \quad g(x) \in U, \qquad (-\infty, \infty);$$

$$2. \quad a(x) \in B, \qquad (-\infty, \infty);$$

$$3. \quad h(x) \in L, \qquad (-\infty, \infty);$$

$$4. \quad \int_{-\infty}^{\infty} g(x-t)a(t)\, dt \to A \int_{-\infty}^{\infty} g(t)\, dt, \qquad x \to \infty$$

$$\Rightarrow \quad \int_{-\infty}^{\infty} h(x-t)a(t)\, dt \to A \int_{-\infty}^{\infty} h(t)\, dt, \qquad x \to \infty.$$

By considering the function $a(t) - A$ we see that there is no loss of generality in assuming $A = 0$. Set

$$G = g * a, \qquad H = h * a.$$

From the assumption $G(\infty) = 0$ we wish to prove $H(\infty) = 0$. Assuming the contrary, there must exist $\delta > 0$ and a sequence x_n tending to $+\infty$ with n such that $|H(x_n)| > \delta$. Now observe that

$$(3.2) \qquad G * h = H * g = g * h * a.$$

This follows by Fubini's theorem in the presence of hypotheses 1, 2, 3.

Now consider the sequence

$$s_n(x) = H(x + x_n).$$

We shall show that it is bounded and equicontinuous on $-\infty < x < \infty$. For, if M is an upper bound for $|a(t)|$, then

$$|H(x)| \leq M \int_{-\infty}^{\infty} |h(t)| \, dt, \qquad -\infty < x < \infty,$$

$$|s_n(x) - s_n(y)| \leq M \int_{-\infty}^{\infty} |h(x + x_n - t) - h(y + x_n - t)| \, dt = o(1),$$

$$y - x \to 0.$$

The latter result follows from a classical result of Lebesgue theory; see N. Wiener [1933, p. 14].

By Ascoli's lemma (see below) we now select from the functions $s_n(x)$ a subset $s_{n_k}(x)$ which tends to a limit $s(x)$, continuous on $(-\infty, \infty)$. By Lebesgue's limit theorem, applicable since $H \in B$, $g \in L$, we have

$$\lim_{k \to \infty} \int_{-\infty}^{\infty} s_{n_k}(x - t)g(t) \, dt = \lim_{k \to \infty} \int_{-\infty}^{\infty} H(x + x_{n_k} - t)g(t) \, dt$$

$$(3.3) \quad = \lim_{k \to \infty} \int_{-\infty}^{\infty} G(x + x_{n_k} - t)h(t) \, dt = \int_{-\infty}^{\infty} s(x - t)g(t) \, dt = g * s.$$

Here we have used equation (3.2). Since $G(\infty) = 0$, another application of the Lebesgue limit theorem shows that the limit (3.3) is zero. Since $s(x) \in \text{B.C}$ we infer from hypothesis 1 and Definition 3.2 that $s(x) \equiv 0$. But

$$|s(0)| = \lim_{k \to \infty} |s_{n_k}(0)| = \lim_{k \to \infty} |H(x_{n_k})| \geq \delta > 0.$$

The contradiction shows that $H(\infty) = 0$, as we wished to prove.

For the reader's convenience we prove the form of Ascoli's lemma which we have used.

Lemma 3.1.

1. $\{s_n(x)\}_1^{\infty}$ is equicontinuous on $-\infty < x < \infty$;

2. $|s_n(x)| < M, \qquad n = 1, 2, \ldots, \qquad -\infty < x < \infty, \qquad$ some M

$\Rightarrow$ A. $\lim_{k \to \infty} s_{n_k}(x) = s(x),$ some integers n_k, some $s(x)$;

     B. $s(x) \in C, \qquad -\infty < x < \infty.$

Arrange all real rational numbers in a sequence $x_1, x_2, x_3, \ldots$. From the bounded double sequence $s_n(x_m)$ we may pick, by the familiar diagonal process, a subsequence $s_{n_k}(x_m)$ such that

$$(3.4) \qquad \lim_{k \to \infty} s_{n_k}(x_m) = s(x_m), \qquad m = 1, 2, 3, \ldots .$$

See, for example, D. V. Widder [1946, p. 26]. The function $s(x)$ is defined by the limit (3.4) at all rational points.

Now let y be irrational and let $\varepsilon > 0$ be arbitrary. By hypothesis 1, if x_m is sufficiently near to y,

$$|s_{n_k}(y) - s_{n_k}(x_m)| < \varepsilon, \qquad k = 1, 2, 3, \ldots ,$$
$$|s_{n_j}(y) - s_{n_j}(x_m)| < \varepsilon, \qquad j = 1, 2, 3, \ldots .$$

Such an x_m exists since y is a limit point of rational points. By Cauchy's limit criterion equation (3.4) implies the existence of a number K such that

$$|s_{n_k}(x_m) - s_{n_j}(x_m)| < \varepsilon, \qquad k > K, \quad j > K.$$

By the triangle inequality

$$|s_{n_k}(y) - s_{n_j}(y)| < 3\varepsilon, \qquad k > K, \quad j > K,$$

from which we may infer the existence of

$$(3.5) \qquad \lim_{k \to \infty} s_{n_k}(y) = s(y).$$

It remains to show only that $s(x)$, now defined for all x, is continuous. By the equicontinuity of the sequence $s_{n_k}(x)$ we have for a given ε the existence of a number δ, independent of n_k, such that

$$|s_{n_k}(x) - s_{n_k}(y)| < \varepsilon, \qquad |x - y| < \delta,$$
$$|s(x) - s(y)| \leqq \varepsilon, \qquad |x - y| < \delta.$$

Thus the lemma is proved.

We point out that Ascoli's result is usually stated for compact regions, in which case the limit (3.5) is uniform. That the approach need not be uniform on $(-\infty, \infty)$ may be seen by considering $s_n(x) = \sin(x/n)$. Here $s(x) \equiv 0$, but any inequality of the form $|s_n(x)| < \varepsilon < 1$ must fail (at $x = n\pi/2$, for example).

The previous theorem, restated for the interval $(0, \infty)$ becomes the following.

Theorem 3.2.

1. $g(x) \in U$, $(0, \infty)$;

2. $a(x) \in B$, $(0, \infty)$;

3. $h(x) \in L$, $(0, \infty)$;

4. $\dfrac{1}{x} \displaystyle\int_0^\infty g\!\left(\dfrac{t}{x}\right) a(t)\, dt \to A \int_0^\infty g(t)\, dt$, $x \to \infty$

$\Rightarrow \qquad \dfrac{1}{x} \displaystyle\int_0^\infty h\!\left(\dfrac{t}{x}\right) a(t)\, dt \to A \int_0^\infty h(t)\, dt$, $x \to \infty$.

We may specialize this result by choosing $g(x) = e^{-x} x^{\alpha-1}$ and $h(x) = x^{\beta-1}$ on $(0, 1)$, $h(x) = 0$ on $(1, \infty)$. The hypotheses 1 and 3 are evidently satisfied if $\alpha > 0$, $\beta > 0$.

Theorem 3.3

1. $f(x) = \displaystyle\int_0^\infty e^{-xt} t^{\alpha-1} a(t)\, dt$, $\alpha > 0$;

2. $a(x) \in B$, $(0, \infty)$;

3. $f(x) \sim \dfrac{A\Gamma(\alpha)}{x^\alpha}$, $x \to 0+$

$\Rightarrow \qquad \displaystyle\int_0^x t^{\beta-1} a(t)\, dt \sim \dfrac{Ax^\beta}{\beta}$, $\beta > 0$, $x \to \infty$.

In particular if $\alpha = \beta = 1$, this result reduces to Theorem 2.4, the proof of which we had deferred.

4. Hardy's and Littlewood's Integral Tauberian Theorems

As we noted in §1, Hardy proved first that if $a_k = O(1/k)$, $k \to \infty$, then the series (1.1) cannot be summable in the sense of Cesàro unless it converges; Littlewood (1912, p. 434) proved the same result for Abel summability. In this section we shall prove the integral analogs of these results, deriving therefrom in §7 the classical series result. We need first a new definition.

Definition 4. A function $f(x) \in \mathrm{SO}$ (slowly oscillating) on $(0, \infty)$ if

$$\lim [f(y) - f(x)] = 0$$

when y and $x \to \infty$ in such a way that $y/x \to 1$.

For example, if $f(x) \in C^1$ and $x|f'(x)| < M$ on $(0, \infty)$ then $f(x) \in \mathrm{SO}$ there. For,

$$|f(y) - f(x)| \leq \left| \int_x^y f'(t)\, dt \right| \leq M \log\left(\frac{y}{x}\right) = o(1), \qquad \frac{y}{x} \to 1.$$

Thus $\sin (\log x) \in \mathrm{SO}$ on $(0, \infty)$. On the other hand, $\sin x \notin \mathrm{SO}$ on $(0, \infty)$. (See Exercise 6 of this chapter.)

Theorem 4.1.

$$1. \quad a(x) \in \mathrm{SO}, \qquad (0, \infty);$$

$$2. \quad \int_0^x a(t)\, dt \sim Ax, \qquad x \to \infty$$

$$\Rightarrow \qquad a(\infty) = A.$$

As usual, we may assume $A = 0$. From hypothesis 1 we see that to an arbitrary $\varepsilon > 0$ there correspond numbers K and δ such that

$$|a(y) - a(x)| < \varepsilon, \qquad y > K, \quad x > K, \quad \left| \frac{y}{x} - 1 \right| < \delta.$$

That is, for $x - \delta x < y < x + \delta x$

$$a(x) - \varepsilon < a(y) < a(x) + \varepsilon.$$

Integration gives

$$(4.1) \qquad [a(x) - \varepsilon] 2\delta x < \int_{x(1-\delta)}^{x(1+\delta)} a(y)\, dy < [a(x) + \varepsilon]\, 2\delta x.$$

Now let $x \to \infty$. The integral (4.1) is $o(x)$, $x \to \infty$, by hypothesis 2. Thus

$$\overline{\lim}_{x \to \infty} [a(x) - \varepsilon]\, 2\delta \leq 0, \qquad \underline{\lim} [a(x) + \varepsilon]\, 2\delta \geq 0,$$

$$-\varepsilon \leq \underline{\lim}\, a(x) \leq \overline{\lim}\, a(x) \leq \varepsilon.$$

The conclusion is now immediate.

Hardy's integral Tauberian theorem is an immediate consequence of this result.

Theorem 4.2.

$$1. \quad xa(x) \in B, \qquad (0, \infty);$$

$$2. \quad A(x) = \int_0^x a(t)\, dt;$$

$$3. \quad \int_0^x A(t)\, dt \sim Ax, \qquad x \to \infty$$

$$(4.2) \quad \Rightarrow \qquad \int_0^\infty a(t)\, dt = A.$$

Hypothesis 1 corresponds to condition C of §1 and hypothesis 3 states that the integral (4.2) is summable (C) to the value A. We apply Theorem 4.1 to the function $A(x)$. It belongs to SO since

$$|a(y) - a(x)| \leqq M \left| \int_x^y \frac{dt}{t} \right| = M \left| \log \frac{y}{x} \right| = o(1), \qquad \frac{y}{x} \to 1.$$

Here M is a bound for $|xa(x)|$. The conclusion of Theorem 4.1 is $A(\infty) = A$, equivalent to (4.2).

The proof of Littlewood's integral Tauberian theorem is slightly more complicated.

Theorem 4.3.

$$1. \quad xa(x) \in B, \qquad (0, \infty);$$

$$2. \quad f(x) = \int_0^\infty e^{-xt} a(t)\, dt \to A, \qquad x \to 0+$$

$$(4.3) \quad \Rightarrow \qquad \int_0^\infty a(t) = A.$$

By use of Theorem 2.4 we show first that, in the presence of hypothesis 1, the Abel summability of the integral (4.3) implies its Cesàro summability, and we can then apply Theorem 4.2. An obvious integration by parts enables us to write hypothesis 2 as

$$\int_0^\infty e^{-xt} A(t)\, dt \sim Ax, \qquad x \to 0+.$$

To prove that $A(x) \in B$, consider the difference

$$A(x) - f\!\left(\frac{1}{x}\right) = \int_0^x a(t)[1 - e^{-t/x}]\, dt - \int_x^\infty e^{-t/x} a(t)\, dt.$$

By hypothesis 1

$$\left| A(x) - f\!\left(\frac{1}{x}\right) \right| \le M \int_0^x \frac{1}{x}\, dt + M \int_x^\infty \frac{e^{-t/x}}{t}\, dt$$

$$\le M + M \int_1^\infty \frac{e^{-t}}{t}\, dt \le 2M.$$

Since $f(x) \in B$ on $(0, \infty)$ by hypothesis 2, the same must be true of $A(x)$. Hence if we apply Theorem 2.4 to $A(x)$ we find that

$$\int_0^x A(t)\, dt \sim Ax, \qquad x \to \infty.$$

That is, the integral (4.3) is summable (C) to A, and the convergence of (4.3) now follows from Hardy's theorem.

As an illustration take $a(t) = (\sin t)/t$. Its Laplace transform is $\tan^{-1}(1/x)$ which tends to $\pi/2$ as $x \to 0+$. Obviously $ta(t) \in B$, so that

$$\int_0^\infty \frac{\sin t}{t}\, dt = \frac{\pi}{2}.$$

5. One-Sided Tauberian Conditions

In Hardy's and Littlewood's theorems the condition $a(x)x \in B$ can be replaced by a one-sided condition in which $a(x)x$ is bounded below (or above) only. With a view to proving this we first establish several results of J. Karamata [1931, p. 27].

Theorem 5.1.

    1. $g(x) = 0, \quad 0 \leqq x < c < 1; \qquad g(x) \in C, \quad c \leqq x \leqq 1;$

    2. $\varepsilon > 0, \quad \gamma > 0$

$\Rightarrow$ There exist polynomials $p(x)$, $P(x)$ such that

(5.1) $$p(x) < g(x) < P(x), \qquad 0 \leqq x \leqq 1,$$

(5.2) $$\int_0^1 \left[\log\left(\frac{1}{x}\right)\right]^{\gamma-1} [P(x) - p(x)]\, dx < \varepsilon.$$

For an arbitrary positive number η we can determine $h(x) \in C$ on $0 \leqq x \leqq 1$ such that

$$g(x) \leqq h(x), \qquad 0 \leqq x \leqq 1.$$

(5.3) $$\int_0^1 \left[\log\left(\frac{1}{x}\right)\right]^{\gamma-1} [h(x) - g(x)]\, dx < \eta.$$

For example, if $g(c+) > 0$ (the only case needed later) we may take $h(x) = g(x)$ except in a small interval $(c - \delta, c)$ where $h(x) = g(c+)(x - c + \delta)/\delta$. The value of the integral (5.3) is less than

$$g(c+) \int_{c-\delta}^c \left[\log\left(\frac{1}{x}\right)\right]^{\gamma-1} dx = o(1), \qquad \delta \to 0.$$

Hence inequality (5.3) may be satisfied by choice of δ. By the Weierstrass approximation theorem we determine a polynomial $Q(x)$ such that $|h(x) - Q(x)| < \eta$ on $0 \leqq x \leqq 1$, and we define $P(x) = Q(x) + \eta$. Then $g(x) \leqq h(x) < P(x)$ and

(5.4) $$\int_0^1 \left[\log\left(\frac{1}{x}\right)\right]^{\gamma-1} (P - g)\, dx \leqq \int_0^1 \left[\log\left(\frac{1}{x}\right)\right]^{\gamma-1}$$

$$\times [(P - Q) + |Q - h| + (h - g)]\, dx < \eta\Gamma(\gamma) + \eta\Gamma(\gamma) + \eta.$$

Similarly, there is a polynomial $p(x)$ such that $p(x) < g(x)$ on $(0, 1)$ and such that

(5.5) $$\int_0^1 \left[\log\left(\frac{1}{x}\right)\right]^{\gamma-1} (g - p)\, dx \leqq 2\eta\Gamma(\gamma) + \eta.$$

Adding inequalities (5.4) and (5.5) and choosing $\eta = \varepsilon/(4\Gamma(\gamma) + 2)$, we have (5.2). Hypothesis 1 could be very considerably relaxed if desired; see D. V. Widder [1941, p. 189].

Theorem 5.2.

1. $\quad f(x) = \displaystyle\int_0^\infty e^{-xt} a(t)\, dt, \qquad 0 < x < \infty;$

2. $\quad a(x) \geqq 0, \qquad 0 < x < \infty;$

3. $\quad g(x) = 0, \quad 0 \leqq x < c < 1; \qquad g(x) \in C, \quad c \leqq x \leqq 1;$

4. $\quad f(x) \sim \dfrac{A}{x^\gamma}, \qquad \gamma > 0, \quad x \to 0+.$

$$(5.6) \quad \Rightarrow \quad \int_0^\infty e^{-tx} g(e^{-tx}) a(t)\, dt \sim \frac{A}{\Gamma(\gamma)}\frac{1}{x^\gamma}\int_0^\infty e^{-t} t^{\gamma-1} g(e^{-t})\, dt,$$
$$x \to 0+.$$

We begin by determining the polynomials p and P of Theorem 5.1, corresponding to a given ε. Since $a(x) \geqq 0$ we have

(5.7)
$$\int_0^\infty e^{-xt} p(e^{-xt}) a(t)\, dt \leqq \int_0^\infty e^{-xt} g(e^{-xt}) a(t)\, dt \leqq \int_0^\infty e^{-xt} P(e^{-xt}) a(t)\, dt,$$

(5.8)
$$\int_0^\infty e^{-t} p(e^{-t}) t^{\gamma-1}\, dt \leqq \int_0^\infty e^{-t} g(e^{-t}) t^{\gamma-1}\, dt \leqq \int_0^\infty e^{-t} P(e^{-t}) t^{\gamma-1}\, dt.$$

If we replace x by $(n + 1)x$ in hypothesis 4 we obtain

$$(5.9) \quad \int_0^\infty e^{-xt} e^{-nxt} a(t)\, dt \sim \frac{A}{(n+1)^\gamma}\frac{1}{x^\gamma} = \frac{A}{x^\gamma}\frac{1}{\Gamma(\gamma)}\int_0^\infty e^{-t} e^{-nt} t^{\gamma-1}\, dt.$$

Thus the conclusion (5.6) is valid if $g(x)$ is replaced by x^n and hence by any polynomial. From (5.7) and (5.9) we have

$$A\int_0^\infty e^{-t} p(e^{-t}) t^{\gamma-1}\, dt \leqq \varliminf_{x \to 0+} x^\gamma \Gamma(\gamma) \int_0^\infty e^{-xt} g(e^{-xt}) a(t)\, dt$$

$$\leqq \varlimsup_{x \to 0+} x^\gamma \Gamma(\gamma) \int_0^\infty e^{-xt} g(e^{-xt}) a(t)\, dt \leqq A\int_0^\infty e^{-t} P(e^{-t}) t^{\gamma-1}\, dt.$$

By (5.2) the two extremes of this inequality differ by $A\varepsilon$. Hence the two terms in the middle must be equal to each other and to the middle term of (5.8). That is, (5.6) is established.

We now specialize the function $g(x)$ to obtain the following Tauberian theorem.

Theorem 5.3.

$$1. \quad f(x) = \int_0^\infty e^{-xt} a(t)\, dt, \qquad 0 < x < \infty;$$

$$2. \quad a(x) \geqq 0, \qquad 0 < x < \infty;$$

$$3. \quad f(x) \sim \frac{A}{x^\gamma}, \qquad \gamma > 0, \quad x \to 0+$$

$$(5.10) \quad \Rightarrow \qquad \int_0^x a(t)\, dt \sim \frac{A x^\gamma}{\Gamma(\gamma + 1)}, \qquad x \to \infty.$$

Except for the present one-sided Tauberian condition 2 this result would be included in Theorem 3.3. To prove it, choose $g(x) = 1/x$ on $(1/e, 1)$, $g(x) = 0$ on $(0, 1/e)$ in Theorem 5.2. Then (5.6) becomes

$$\int_0^{1/x} a(t)\, dt \sim \frac{A}{x^\gamma \Gamma(\gamma)} \int_0^1 t^{\gamma - 1}\, dt, \qquad x \to 0+,$$

which is equivalent to (5.10).

6. One-sided Version of Littlewood's Integral Theorem

To obtain a one-sided version of Theorem 4.3 we need several preliminary results, of interest in themselves. We introduce a definition.

Definition 6. A function $f(x) \in \mathrm{SD}$ (slowly decreasing) on $(0, \infty)$ if

$$(6.1) \qquad \varliminf \, [f(y) - f(x)] \geqq 0,$$

when y and x become infinite in such a way that $y > x$ and $y/x \to 1$.

For example, any increasing function belongs to this class. Or, $f(x) \in SD$ if $xf'(x) > -M$ for some constant M. For,

$$f(y) - f(x) > -M \log\left(\frac{y}{x}\right), \qquad y > x,$$

from which (6.1) is immediate.

We now prove a one-sided version of Theorem 4.1.

Theorem 6.1.

$$1. \quad a(x) \in SD, \qquad (0, \infty);$$

$$2. \quad \int_0^x a(t)\,dt \sim Ax, \qquad x \to \infty$$

$$\Rightarrow \qquad a(\infty) = A.$$

Assume $A = 0$. For an arbitrary $\varepsilon > 0$ we have from (6.1) that

$$(6.2) \qquad a(x) - \varepsilon < a(y), \qquad x < y < (1 + \delta)x.$$

Integrating with respect to y and using hypothesis 2, we obtain

$$[a(x) - \varepsilon]\delta x < \int_x^{x(1+\delta)} a(y)\,dy = o(x), \qquad x \to \infty,$$

$$(6.3) \qquad \overline{\lim_{x \to \infty}} \; a(x) \leqq \varepsilon.$$

Rewrite (6.2) as

$$a(x) < a(y) + \varepsilon, \qquad \frac{y}{1 + \delta} < x < y$$

and integrate with respect to x:

$$o(y) < [a(y) + \varepsilon]\frac{\delta y}{1 + \delta}, \qquad y \to \infty, \qquad \underline{\lim_{y \to \infty}} \; a(y) \geqq -\varepsilon.$$

This with (6.3) shows that $a(\infty) = 0$, as we wished to prove.

Corollary 6.1.

$$1. \quad xa(x) > -M, \qquad \text{some } M, \quad 0 < x < \infty;$$

$$2. \quad \int_0^\infty a(x)\,dx \quad \text{is summable (C) to } A$$

$$\Rightarrow \qquad \int_0^\infty a(x)dx = A.$$

This is a one-sided version of Theorem 4.2. It follows at once by observing that $A(x)$ now satisfies the conditions of Theorem 6.1.

We prove next a result of E. Landau [1929, p. 58].

Theorem 6.2.

$$1. \quad f(x) \in C^2, \qquad 0 \leqq x < \infty;$$

$$2. \quad f(0) = A;$$

$$3. \quad f''(x) > \frac{-M}{x^2}, \qquad \text{some } M, \quad 0 < x < \infty$$

$$\Rightarrow \qquad f'(x) = o\left(\frac{1}{x}\right), \qquad x \to 0+.$$

From Taylor's theorem we have for $0 < \delta < 1$

$$f(x \pm \delta x) - f(x) \mp \delta x \, f'(x) = \frac{\delta^2 x^2}{2} f''(x \pm \theta \delta x), \qquad 0 < \theta < 1$$

$$> \frac{-M\delta^2}{2(1 \pm \theta \delta)^2}.$$

Now let $x \to 0+$, noting that $f(x + \delta x) - f(x) \to 0$ by hypothesis 2. We obtain

$$\frac{-M\delta}{2(1 - \delta)^2} \leqq \varliminf_{x \to 0+} xf'(x) \leqq \varlimsup_{x \to 0+} xf'(x) \leqq \frac{M\delta}{2}.$$

Since δ is arbitrary we obtain at once the desired conclusion.

We derive next a consequence of Theorem 4.3, using a modification of the first hypothesis thereof.

Theorem 6.3.

$$1. \quad B(x) = \int_0^x ta(t) \, dt = o(x), \qquad x \to \infty;$$

$$2. \quad f(x) = \int_0^\infty e^{-xt} a(t) \, dt \to A, \qquad x \to 0+$$

$$(6.4) \qquad \Rightarrow \qquad \int_0^\infty a(t) \, dt = A.$$

This states that hypothesis 1 is an alternative condition to Littlewood's for the Abel summability of the integral (6.4) to imply its convergence. Assume $A = 0$. Integration by parts gives

$$(6.5) \quad \int_1^\infty e^{-xt} \frac{B(t)}{t^2}\, dt = B(1)e^{-x} - x \int_1^\infty e^{-xt} \frac{B(t)}{t}\, dt + \int_1^\infty e^{-xt} a(t)\, dt.$$

The determining function $B(t)/t^2$ of this first integral is $O(1/t)$ by hypothesis 1. Hence we may apply Theorem 4.3. Its limit as $x \to 0+$, from the right-hand side of (6.5) is

$$B(1) - \int_0^1 a(t)\, dt.$$

For,

$$\lim_{x \to 0+} x \int_1^\infty e^{-xt} \frac{B(t)}{t}\, dt = 0$$

by Theorem 2.1, and

$$\lim \int_1^\infty e^{-xt} a(t)\, dt = - \lim \int_0^1 e^{-xt} a(t)\, dt = - \int_0^1 a(t)\, dt$$

by hypothesis 2. Thus the conclusion of Theorem 4.3 yields

$$(6.6) \qquad \int_1^\infty \frac{B(t)}{t^2}\, dt = B(1) - \int_0^1 a(t)\, dt,$$

$$(6.7) \qquad \qquad = B(1) + \int_1^\infty a(t)\, dt.$$

To obtain (6.7) we have integrated integral (6.6) by parts. Equation (6.4), with $A = 0$, is now immediate by comparing (6.6) with (6.7).

We can now obtain the desired one-sided version of Littlewood's theorem very easily.

Theorem 6.4.

$$1. \quad a(x) > \frac{-M}{x}, \qquad \text{some } M, \quad 0 < x < \infty;$$

$$2. \quad f(x) = \int_0^\infty e^{-xt} a(t)\, dt \to A, \qquad x \to 0+$$

$$\Rightarrow \qquad \int_0^\infty a(t)\, dt = A.$$

From hypothesis 1 we have

$$f''(x) = \int_0^\infty e^{-xt} t^2 a(t)\, dt > \frac{-M}{x^2}, \qquad 0 < x < \infty.$$

By Theorem 6.2 $f'(x) = o(1/x)$, $x \to 0+$. Thus

$$\int_0^\infty e^{-xt}[M + ta(t)]\, dt = \frac{M}{x} - f'(x) \sim \frac{M}{x}, \qquad x \to 0+.$$

Since $M + ta(t) > 0$ we may apply Theorem 5.3 to this integral to obtain

$$\int_0^x [M + ta(t)]\, dt \sim Mx, \qquad x \to \infty.$$

This implies that hypothesis 1 of Theorem 6.3 is valid. The conclusion of that theorem is the desired result here.

7. Classical Series Results

We have hitherto proved the integral analogs of the Tauberian theorems about series mention in §1. It is now possible to obtain these discrete results, not by analogy, but as direct consequence of the continuous cases. We do so in a few instances, using the more general one-sided versions. First, the discrete analog of Theorem 2.4 or of Theorem 5.3 ($\gamma = 1$):

Theorem 7.1.

$$(7.1) \qquad 1. \quad \sum_{k=0}^\infty a_k e^{-kx} \to A, \qquad x \to 0+ ;$$

$$2. \quad -M < s_n = \sum_{k=0}^n a_k, \qquad \text{some } M, \quad n = 0, 1, 2, \ldots$$

$$\Rightarrow \qquad \sum_{k=0}^n s_k \sim A(n + 1), \qquad n \to \infty.$$

That is, if the sequence of partial sums of a series is bounded on one side it cannot be summable in the sense of Abel unless it is also summable in the sense of Cesàro. It is obvious that for $\gamma = 1$ in Theorem 5.3 the lower bound zero of hypothesis 2 may be replaced by any number $-M$. (See Exercise 15 of this chapter.)

Define a function $A(x)$ so that $A(0) = 0$ and

$$A(x) = \sum_{k=0}^{[x]} a_k > -M, \qquad x > 0.$$

If the sum of the series (7.1) is $f(x)$, we have

$$f(x) = \int_0^\infty e^{-xt}\, dA(t) = x \int_0^\infty e^{-xt} A(t)\, dt, \qquad x > 0.$$

Thus by hypothesis 1

$$\int_0^\infty e^{-xt} A(t)\, dt \sim \frac{A}{x}, \qquad x \to 0+.$$

The conclusion of Theorem 5.3, with x replaced by $n + 1$, is

$$\int_0^{n+1} A(t)\, dt = \sum_{k=0}^{n} s_k \sim A(n + 1), \qquad n \to \infty.$$

This is the desired result.

We turn next to the series form of Hardy's theorem.

Theorem 7.2.

$$\text{1.} \quad a_k > \frac{-M}{k}, \qquad \text{some } M, \quad k = 1, 2, \ldots\,;$$

$$\text{2.} \quad s_n = \sum_{k=0}^{n} a_k, \qquad \sum_{k=0}^{n} s_k \sim A(n + 1), \quad n \to \infty$$

$$\Rightarrow \qquad \sum_{k=0}^{\infty} a_k = A.$$

Observe first that hypothesis 2 implies that $s_n = o(n)$ and hence that $a_n = o(n)$, $n \to \infty$. Define $A(x)$ as in the previous theorem and apply Theorem 6.1 to it. Set $m = [y]$ and $n = [x]$, $y > x$. Then

$$A(y) - A(x) = \sum_{n+1}^{m} a_k > -M \sum_{n+1}^{m} \frac{1}{k} > -M \int_n^m \frac{dt}{t}, \qquad M > 0.$$

From these inequalities it is clear that $A(x) \in \text{SD}$ on $(0, \infty)$. Further, if $n + 1 \leq x < n + 2$

$$\int_0^x A(t)\, dt = \sum_{k=0}^n s_k + a_{n+1}(x - n - 1),$$

$$o(1) + \frac{1}{n+2} \sum_{k=0}^n s_k < \frac{1}{x} \int_0^x A(t)\, dt \leq \frac{1}{n+1} \sum_{k=0}^n s_k + o(1), \qquad n \to \infty.$$

Then by hypothesis 2

$$\int_0^x A(t)\, dt \sim Ax, \qquad x \to \infty.$$

The conclusion of Theorem 6.1 is

$$A(\infty) = \sum_{k=0}^\infty a_k = A,$$

as we wished to prove.

The series version of Littlewood's theorem follows.

Theorem 7.3.

$$1. \quad a_k > \frac{-M}{k+1}, \qquad \text{some } M, \quad k = 0, 1, 2, \ldots;$$

$$2. \quad \sum_{k=0}^\infty a_k e^{-kx} \to A, \qquad x \to 0+$$

$$\Rightarrow \qquad \sum_{k=0}^\infty a_k = A.$$

We apply Theorem 6.4 defining $a(x)$ by

$$a(x) = a_k, \qquad k \leq x < k + 1, \qquad k = 0, 1, 2, \ldots.$$

From hypothesis 1 it is clear that $a(x) > -M/x$ for $1 \leq x < \infty$. The conclusion of Theorem 6.4 is

$$\int_0^\infty a(t)\, dt = \sum_{k=0}^\infty a_k = A,$$

as we wished to prove.

8. Summary

A typical Abelian theorem is the consistency or regularity one which states that a series or integral which converges is also summable, say in Abel's sense. A typical Tauberian theorem is a modified converse, say Littlewood's, to the effect that the integral $\int_0^\infty a(t)\,dt$, where $a(t) = O(1/t)$, $t \to \infty$, cannot be summable, in Abel's sense, unless it converges. In proving this and similar results we used two approaches. In the first, a modification of Wiener's method, we used integrals with very general kernels but imposed Tauberian conditions that were "two-sided"; in the second we used the Laplace kernel e^{-xt} only but imposed the less restrictive "one-sided" conditions. The results thus overlap in generality. It is noteworthy that no appeal to the theory of Fourier integrals was needed, nothing more recondite having been used than Ascoli's selection theorem in the first method and Weierstrass's approximation theorem for the second.

EXERCISES

1. Show that Cesàro summability, as defined by (1.3) for series and by (2.3) for integrals, is regular.

2. Illustrate Theorem 2.3 by the example

$$a(t) = 1 - \cos t, \quad f(x) = 1/(x^3 + x).$$

3. Use the example $a(t) = 1 + t \sin t$ to show that Theorem 2.4 would be false without hypothesis 2.

4. Prove the power series analog of Theorem 2.4 by use of that theorem:

    1. $|a_k| < M,$ some M, $k = 0, 1, 2, \ldots.$

    2. $\displaystyle\sum_{k=0}^{\infty} a_k x^k \sim \frac{A}{1-x},$ $x \to 1-.$

$$\Rightarrow \qquad \sum_{k=0}^{n} a_k \sim An, \qquad n \to \infty.$$

5. Use Theorem 2.1 to prove that

$$a(\infty) = A \Rightarrow \int_0^\infty e^{-xt} a(t)\, dt \sim \frac{A}{x}, \qquad x \to 0+.$$

6. Prove that $\sin x \notin$ SO on $(0, \infty)$.
[Hint: $\sin(x + \pi) - \sin x = -2 \sin x$.]

7. Prove that $f(x) \in$ SD on $(0, \infty)$ if $f(x) \in B$, $xf(x) \in \uparrow$.

$$\left[\text{Hint: } f(y) - f(x) = \left[f(y) - f(x) \frac{x}{y} \right] + f(x)\left(\frac{x}{y} - 1 \right) \right.$$

$$\left. \geq f(x)\left(\frac{x}{y} - 1 \right). \right]$$

8. Show that $ax + b \cos x \in$ SD on $(0, \infty)$. if $a > 0$

9. Show that $f(x) = x + \cos x + \cos(\log x) \in$ SD on $(0, \infty)$.

10. Show that $f(x)$ in Exercise 9 is such that $xf(x) \notin \uparrow$ on $(0, \infty)$.

$$\left[\text{Hint: } (xf(x))' = x(2 - \sin x) + \cos x + \sqrt{2} \sin\left(\frac{\pi}{4} - \log x \right). \right]$$

11. Use $f(x)$ of Exercise 9 to determine whether Lemma 5.5 of Chapter 4 includes the present Theorem 6.1.

12. If $h(t) = 1$ on $(1, e)$, $h(t) = 0$ elsewhere, show $h(t) \notin U$ on $(0, \infty)$.

$$\left[\text{Hint: Take } a(t) = \frac{d}{dt} \sin(2 \pi \log t), \text{ for example.} \right]$$

13. By integrating formula 3, §2 of Chapter 1, prove rigorously that

$$(9.1) \qquad \tan^{-1}\left(\frac{1}{x} \right) = \int_0^\infty e^{-xt} \frac{\sin t}{t}\, dt, \qquad x > 0.$$

14. Prove that

$$\frac{1}{x} \tan^{-1}\left(\frac{1}{x} \right) = \int_0^\infty e^{-xt}\, dt \int_0^t \frac{\sin y}{y}\, dy, \qquad x > 0.$$

[Hint: Integrate (9.1) by parts.]

15. Show that Theorem 5.3 remains true if $\gamma \geq 1$ and hypothesis 2 is replaced by $a(x) \geq - M$ on $(0, \infty)$.

16. Show that if M is replaced by 0 in hypothesis 1 of Corollary 6.1 the conclusion follows trivially.

$$\left[\text{Hint:} \frac{1}{x} \int_0^x A(t)\, dt \geqq \frac{1}{x} \int_R^x A(t)\, dt \geqq A(R)\left(1 - \frac{R}{x}\right), \qquad A \geqq A(R). \right]$$

17. Admitting equation (6.7) and Theorem 4.10 of Chapter 4, complete the proof of the prime number theorem by use of Exercise 7 and Theorem 6.1 of the present chapter.

9 *Inversion by Series*

---

1. Introduction

In this chapter we shall treat two special transforms, both of which may be written as

$$(1.1) \qquad f(x) = \int_0^\infty K\left(\frac{x}{t}\right) \frac{\phi(t)}{t}\, dt$$

with suitable kernels $K(t)$. By exponential change of variable, (1.1) becomes the convolution transform

$$f(e^{-x}) = \int_{-\infty}^\infty K(e^{-x+t})\phi(e^{-t})\, dt,$$

whose inversion function $E(s)$ is defined by

$$(1.2) \qquad \frac{1}{E(s)} = \int_{-\infty}^\infty e^{-st}K(e^{-t})\, dt = \int_0^\infty x^{s-1}K(x)\, dx.$$

For both kernels $K(x)$ which we have in mind $E(s)$ will belong to the Laguerre–Pólya class. See §4 of Chapter 7. Consequently, the inversion of (1.1) can then be accomplished by use of the operator $E(D)$, as in

219

Theorem 7.1 of that chapter. However, for these special kernels there is an interesting alternate method of inversion when the generating function $f(x)$ has a power series expansion

$$(1.3) \qquad f(x) = \sum_{k=0}^{\infty} a_k x^k.$$

We shall show that then the determining function $\phi(x)$ also has a power series expansion, which can be obtained from (1.3) by the introduction of a sequence of multipliers:

$$\phi(x) = \sum_{k=0}^{\infty} a_k E(-k) x^k.$$

2. The Potential Transform

The first of the kernels to be treated is

$$K(x) = \frac{2}{\pi} \frac{1}{x^2 + 1}.$$

The transform (1.1) becomes

$$(2.1) \qquad f(x) = \frac{2}{\pi} \int_0^{\infty} \frac{t\phi(t)}{x^2 + t^2} \, dt.$$

The integral is clearly related to the Poisson representation of a harmonic function $u(x, y)$ in the half-plane $y > 0$. We shall make this more explicit later, but on account of this relationship we propose to call (2.1) the *potential transform*.

We can obtain the inversion function $E(s)$ at once from equation (1.2):

$$\frac{1}{E(s)} = \frac{2}{\pi} \int_0^{\infty} \frac{x^{s-1} \, dx}{x^2 + 1} = \frac{1}{\pi} \int_0^{\infty} \frac{y^{(s-2)/2}}{1 + y} \, dy$$

$$= \frac{1}{\pi} B\left(\frac{s}{2}, 1 - \frac{s}{2}\right) = \frac{1}{\sin\left(\dfrac{\pi s}{2}\right)}.$$

Thus $E(s) = \sin(\pi s/2)$ is an entire function with real roots in arithmetic progression. It fails to belong to the class E_0 of §4, Chapter 7, only because $E(0)$ is not equal to unity. This causes no real difficulty because $E(s)/(\pi s/2)$ does belong to $\dot{E}_0$.

The inversion operational equation

$$\text{(2.2)} \qquad \sin\frac{\pi D}{2}\, f(e^{-x}) = \phi(e^{-x})$$

must be interpreted as

$$\text{(2.3)} \qquad \frac{\sin\left(\dfrac{\pi D}{2}\right)}{\left(\dfrac{\pi D}{2}\right)}\left[\frac{\pi D}{2}\, f(e^{-x})\right] = \phi(e^{-x})$$

if we wish to apply Theorem 7.1 of Chapter 7. We shall make this remark more explicit in §5. We observe also that (2.1) can be reduced to the Stieltjes transform:

$$f(\sqrt{x}) = \frac{1}{\pi}\int_0^\infty \frac{\phi(\sqrt{t})}{x+t}\, dt.$$

3. A Brief Table

Let us list here a few pairs (f, ϕ) which satisfy equation (2.1).

	$f(x)$	$\phi(t)$
A.	$\dfrac{1}{x+1}$	$\dfrac{t}{t^2+1}$
B.	e^{-x}	$\sin t$
C.	$\dfrac{e^{-x}}{x}$	$\dfrac{\cos t}{t}$
D.	$\dfrac{\tan^{-1} x}{x}$	$\begin{cases} 0, & 0 < t < 1 \\[2mm] \dfrac{\pi}{(2t)}, & 1 < t < \infty \end{cases}$

E. $\qquad \log\left[\dfrac{x^2+1}{x^2}\right]$ $\qquad \begin{cases} \pi & 0 < t < 1 \\ 0, & 1 < t < \infty \end{cases}$

F. $\qquad \dfrac{\log x}{x^2-1}$ $\qquad \dfrac{\pi}{2}\dfrac{1}{t^2+1}$

G. $\qquad \dfrac{x \log x}{x^2-1}$ $\qquad \dfrac{t \log t}{t^2+1}$

From the integral (2.1) it is clear that the generating function $f(x)$ is always an even function. In the above table it is to be understood that $x > 0$, so that absolute value signs can be omitted.

Let us indicate briefly how this table could be derived. Formula A follows by direct integration of the identity

$$\frac{t^2}{(x^2+t^2)(t^2+1)} = \frac{1}{1-x^2}\left[\frac{1}{t^2+1} - \frac{x^2}{x^2+t^2}\right].$$

Formula 3 of §2, Chapter 1, was

$$\frac{a}{x^2+a^2} = \int_0^\infty e^{-xt}\sin at \, dt.$$

If we regard this as a sine-transform, the inverse is

$$e^{-xt} = \frac{2}{\pi}\int_0^\infty \frac{a}{x^2+a^2}\sin at \, da,$$

from which Formula B above follows by setting $t = 1$. Formula C can be obtained in a similar way from the Laplace transform of cos at. Formulas D and E are familiar ones in the integral calculus.

To obtain Formulas F and G we introduce a parameter s and integrate the following identity with respect to t from 0 to ∞,

$$\frac{t^s}{t^2+1} - \frac{t^s}{x^2+t^2} = \frac{t^s(x^2-1)}{(t^2+1)(x^2+t^2)}.$$

From the familiar formula

$$\frac{2}{\pi}\int_0^\infty \frac{t^s}{x^2+t^2}\,dt = \frac{x^{s-1}}{\cos\dfrac{\pi s}{2}}, \qquad -1 < s < 1,$$

we have

$$(3.1) \qquad \frac{2}{\pi} \int_0^\infty \frac{t^s \, dt}{(t^2 + 1)(x^2 + t^2)} = \frac{1}{x^2 - 1} \left[\frac{1 - x^{s-1}}{\cos \dfrac{\pi s}{2}} \right].$$

The Mellin integral clearly converges for $-1 < s < 3$ and accordingly represents an analytic function of s in that interval. Hence the identity (3.1) must hold there, taking account of the removable singularity at $s = 1$, where the analytic function on the right has the value $(2/\pi)(\log x)/(x^2 - 1)$. This result is equivalent to Formula F. Note that we get a new proof of Formula A by setting $s = 2$ in (3.1).

Finally to obtain Formula G differentiate (3.1) with respect to s and set $s = 2$. The differentiation under the integral sign is valid by Theorem 4.2 of Chapter 5. We obtain

$$\frac{2}{\pi} \int_0^\infty \frac{t^2 \log t \, dt}{(t^2 + 1)(x^2 + t^2)} = \frac{x \log x}{x^2 - 1},$$

as indicated in Formula G.

See Exercise 6-10 of this chapter for alternate derivations of some of these transform pairs.

4. The Inversion Algorithm

In this section let us experiment with an inversion procedure made plausible by the operational equation (2.2). It is applicable, as described below, when the generating function has an expansion for $x > 0$ in a convergent series of powers. Let us describe the conjectured algorithm in three steps:

   I.   Expand the generating function in powers of x.
   II. Cancel the even powers.
   III. Change signs of alternate terms in the resulting series.

The sum of the final series should be the determining function $\phi(x)$. Let us try the process in a few examples.

Example A.

$$\text{I.} \quad \frac{1}{x+1} = 1 - x + x^2 - x^3 + \cdots.$$

$$\text{II.} \quad -x - x^3 - x^5 - x^7 - \cdots.$$

$$\text{III.} \quad x - x^3 + x^5 - x^7 + \cdots = \frac{x}{x^2+1}.$$

Note that the expansion could have been in negative powers of x.

$$\text{I.} \quad \frac{1}{x+1} = \frac{1}{x} - \frac{1}{x^2} + \frac{1}{x^3} - \cdots.$$

$$\text{III.} \quad \frac{1}{x} - \frac{1}{x^3} + \frac{1}{x^5} - \cdots = \frac{x}{x^2+1}.$$

Example B.

$$\text{I.} \quad e^{-x} = 1 - x + \frac{x^2}{2!} - \cdots.$$

$$\text{III.} \quad x - \frac{x^3}{3!} + \frac{x^5}{5!} - \cdots = \sin x.$$

Example C.

$$\text{I.} \quad \frac{e^{-x}}{x} = \frac{1}{x} - 1 + \frac{x}{2!} - \frac{x^2}{3!} + \cdots.$$

$$\text{III.} \quad \frac{1}{x} - \frac{x}{2!} + \frac{x^3}{4!} - \cdots = \frac{\cos x}{x}.$$

In Example D there is an expansion in positive powers of x, but they are all even powers, so that the algorithm leads to $\phi(x) \equiv 0$. This is correct on the interval $(0, 1)$ at least. For Example E the expansion must be in negative even powers of x, and again $\phi(x) \equiv 0$. This too is correct on the interval $(1, \infty)$ at least. Examples F and G have no Maclaurin expansions, and some modification of the algorithm is needed. We shall treat these examples later.

The conjectured inversion procedure seems to have encouraging possibilities. Let us obtain some precise results involving it.

5. The Inversion Operator

Let us alter the symbolic equation (2.2) so as to avoid the exponential change of variable. Since the derivative of $f(e^{-x})$ is $-e^{-x}f'(e^{-x})$ we would do well to define a new operator θ as follows:

$$\theta[f(x)] = -xf'(x).$$

In terms of this new operator equation (2.2) becomes

$$\phi(x) = \sin\frac{\pi\theta}{2}\, f(x) = \frac{\pi}{2}\,\theta\prod_{k=1}^{\infty}\left(1 - \frac{\theta^2}{4k^2}\right)f(x).$$

As in Chapter 7 we now truncate the infinite product to define a linear differential operator of order $2n + 1$ as follows.

Definition 5.

$$L_{n,\,x}[f] = \frac{\pi}{2}\,\theta\prod_{k=1}^{n}\left(1 - \frac{\theta^2}{4k^2}\right)f(x),$$

$$L_x[f] = \lim_{n\to\infty} L_{n,\,x}[f].$$

Note that this operator, when applied to a monomial x^α reproduces it except for a constant factor. For

$$\theta[x^\alpha] = -\alpha x^\alpha.$$

Accordingly, if we define

$$(5.1) \qquad S_n(\alpha) = \frac{\pi}{2}\,\alpha\prod_{k=1}^{n}\left(1 - \frac{\alpha^2}{4k^2}\right),$$

we have

$$L_{n,x}[x^\alpha] = S_n(-\alpha)x^\alpha,$$

or

$$L_x[x^\alpha] = \sin\frac{\pi\theta}{2}\, x^\alpha = -x^\alpha \sin\frac{\pi\alpha}{2}.$$

Since $S_n(\alpha)$ vanishes at the even integers between $-2n$ and $2n$ it is clear that a set of fundamental solutions for the operator $L_{n,x}$ is x^{2k}, $k = 0$, ± 1, $\pm 2,\ldots,\pm n$. Hence $\sin(\pi\theta/2)$ may be described as a linear differential operator which annuls all even powers of x. We illustrate by exhibiting $L_{1,x}$ explicitly:

$$L_{1,n} = \frac{\pi}{8}[x^3 D^3 + 3x^2 D^2 - 3xD].$$

Clearly it annuls the function x^{-2}, 1, x^2 and is an *Euler differential operator* (with the order of each derivative equal to the power of its monomial coefficient), as we might have predicted from the elementary theory of differential equations.

Theorem 5.1.

$$\text{1.} \quad \phi(t) \in \text{C.B}, \qquad 0 < t < \infty;$$

$$(5.2) \quad \text{2.} \quad f(x) = \frac{2}{\pi}\int_0^\infty \frac{t\phi(t)}{x^2 + t^2}\, dt, \qquad \text{converges for } 0 < x < \infty$$

$$(5.3) \quad \Rightarrow \qquad L_x[f] = \phi(x), \qquad 0 < x < \infty.$$

Note that hypothesis 1 alone is not sufficient to guarantee the convergence of the integral (5.2). If $\phi(t)$ is constant, for example, the integral clearly diverges. This is further corroboration that Theorem 7.1 of Chapter 7 is not directly applicable to the present transform. It is equation (2.3) rather than (2.2) which we must use. It becomes, in terms of operator θ,

$$(5.4) \qquad \frac{2}{\pi}\frac{\sin\left(\frac{\pi\theta}{2}\right)}{\theta}\int_0^\infty \frac{2x^2 t}{(x^2 + t^2)^2}\,\phi(t)\, dt = \phi(x).$$

This integral is obtained by applying θ to the integral (5.1), a step which may be justified by showing that the integral (5.4) converges uniformly on (δ, ∞) for an arbitrary positive δ. This is immediate from the inequality

$$\int_0^\infty \frac{t\phi(t)}{(x^2 + t^2)^2}\, dt \ll \underset{(0,\,\infty)}{\text{u.b.}} |\phi(t)| \int_0^\infty \frac{t\, dt}{(\delta^2 + t^2)^2} < \infty, \qquad \delta \leqq x < \infty.$$

If now we apply Theorem 7.1 to the integral (5.4) the conclusion (5.5) follows. As in that theorem we need not assume $\phi(t) \in C$; (5.3) will still follow almost everywhere or at any points of continuity.

We point out here that this theorem can be proved easily without any appeal to the general theory of the convolution transform. It may be shown by induction that

$$(5.5) \quad L_{n,\,x}[f] = \frac{2(2n+1)!}{n!\,n!} \int_0^\infty \frac{x^{2n+2} t^{2n+1}}{(x^2 + t^2)^{2n+2}} \, \phi(t) \, dt, \qquad 0 < x < \infty.$$

See D. V. Widder [1966, p. 357]. Here the integrand is essentially the $(2n + 2)$th power of a function so that Laplace's asymptotic method is applicable (Theorem 2.2 of Chapter 6).

6. Series Inversion

Let us now suppose that in addition to its representation by a potential transform $f(x)$ also has a power series expansion convergent for some positive x.

Theorem 6.1.

$$1. \quad f(x) = \frac{2}{\pi} \int_0^\infty \frac{t\phi(t)}{x^2 + t^2} \, dt, \qquad 0 < x < \infty;$$

$$(6.1) \quad 2. \quad f(x) = \sum_{k=0}^{\infty} a_k x^{k-\alpha}, \qquad 0 < x < \rho, \quad \text{some } \alpha;$$

$$(6.2) \quad \Rightarrow \quad \phi(t) = \sum_{k=0}^{\infty} a_k \sin \frac{\pi}{2} (\alpha - k) \, t^{k-\alpha}, \qquad 0 < t < \delta, \quad \text{some } \delta,$$

at least at points t where $\phi(t) \in C$.

Observe first that the function $S_n(\alpha)$, defined by (5.1) satisfies the inequality

$$|S_n(\alpha)| \leqq \frac{\pi}{2} |\alpha| \prod_{k=1}^{n} \left(1 + \frac{\alpha^2}{4k^2}\right) \leqq \sinh \frac{\pi|\alpha|}{2} \leqq e^{\pi|\alpha|/2}.$$

Now apply the operator $L_{n,t}$ to the series (6.1), a step surely valid for any convergent power series since nothing worse than differentiation is involved:

$$(6.3) \quad L_{n,t}[f] = \sum_{k=0}^{\infty} a_k S_n(\alpha - k) \, t^{k-\alpha}, \qquad 0 < t < \rho.$$

This series converges uniformly in the variable n for all $n > 0$, at least for fixed t in the interval $0 < t < e^{-\pi/2}\rho$. For,

$$\sum_{k=m}^{\infty} a_k S_n(\alpha - k)t^{k-\alpha} \ll \sum_{k=m}^{\infty} |a_k| t^{k-\alpha} e^{\pi(k-\alpha)/2}.$$

The dominant series converges for $0 < t < e^{-\pi/2}\rho$. Here m is any positive number greater than α. By virtue of the uniform convergence we may let n become infinite term by term in series (6.3). The left-hand side tends to $\phi(t)$ by Theorem 5, the right-hand side becomes the series (6.2), and the proof is complete. We may take $\delta = e^{-\pi/2}\rho$. In particular, the present proof shows that if $\rho = \infty$ then $\delta = \infty$.

Note that if $\alpha = 0$ then the multipliers in series (6.2) are

$$- \sin\frac{k\pi}{2}: \qquad 0, -1, 0, 1, 0, -1, 0, \ldots.$$

This verifies the conjectured algorithm of §4.

The corresponding result for expansions in powers of $1/x$ follows.

Theorem 6.2.

$$1. \quad f(x) = \int_0^{\infty} \frac{t\phi(t)}{x^2 + t^2}\, dt, \qquad 0 < x < \infty;$$

$$2. \quad f(x) = \sum_{k=0}^{\infty} a_k x^{-k-\alpha}, \qquad \rho < x < \infty$$

$$\Rightarrow \quad \phi(t) = \sum_{k=0}^{\infty} a_k \sin\frac{\pi}{2}(\alpha + k)\, t^{-k-\alpha}, \qquad R < x < \infty, \quad \text{some } R,$$

at least at points t where $\phi(t) \in C$.

The proof is as above, *mutatis mutandis*. Now $R = e^{\pi/2}\rho$. If $\alpha = 0$ the multipliers now become:

$$\sin\frac{k\pi}{2}: \qquad 0, 1, 0, -1, 0, \ldots.$$

This indicates a difference in the two cases, in the interpretation of the word "alternate" in step III of our algorithm.

In order to handle the inversion of Examples F and G of §3 by the use of series we now indicate that the method is still applicable if the generating function divided by log x has a power series expansion. If we differentiate the equation

$$L_x[x^\alpha] = -\left(\sin\frac{\pi\alpha}{2}\right)x^\alpha$$

with respect to α, thus capitalizing on the linearity of the operator L_x, we obtain

$$(6.4) \qquad L_x[x^\alpha \log x] = -\left(\sin\frac{\pi\alpha}{2}\right)x^\alpha \log x - \frac{\pi}{2}\left(\cos\frac{\pi\alpha}{2}\right)x^\alpha.$$

This relation leads to the following inversion:

$$f(x) = \log x \sum_{k=0}^{\infty} a_k x^k$$

$$(6.5) \quad \Rightarrow L_x[f] = \log x \sum_{k=0}^{\infty} (-1)^{k+1} a_{2k+1} x^{2k+1} + \frac{\pi}{2} \sum_{k=0}^{\infty} (-1)^{k+1} a_{2k} x^{2k},$$

$$f(x) = \log x \sum_{k=0}^{\infty} a_k x^{-k}$$

$$(6.6) \quad \Rightarrow L_x[f] = \log x \sum_{k=0}^{\infty} (-1)^{k} a_{2k+1} x^{-2k-1} + \frac{\pi}{2} \sum_{k=0}^{\infty} (-1)^{k+1} a_{2k} x^{-2k}.$$

The details of proof are similar to those for Theorems 6.1. We omit them here but illustrate one of them by the example

$$f(x) = \frac{\log x}{x^2 - 1} = \log x[-1 - x^2 - x^4 - x^6 - \cdots].$$

$$L_x[f] = \frac{\pi}{2}[1 - x^2 + x^4 - \cdots] = \frac{\pi}{2(1 + x^2)}.$$

This result is verified by Example F of §3.

In Theorem 6.1 one sees by inspection that series (6.2) converges at least as far as series (6.1) does. We have proved equation (6.2) only for a smaller interval, so it is natural to inquire if (6.2) is also valid for $0 < t < \rho$. This is in fact the case, as we now show.

We have seen that the transform (2.1) is equivalent to

$$f(\sqrt{x}) = \frac{1}{\pi}\int_0^\infty \frac{\phi(\sqrt{t})}{x + t}\, dt.$$

From (6.1)

$$\left(\sqrt{x}\right) = \sum_{k=0}^{\infty} a_k x^{(k-\alpha)/2}, \qquad 0 < x < \rho^2.$$

Let us now apply Corollary 16 of Chapter 5:

$$\frac{f(\sqrt{t}\, e^{-i(\pi/2-\varepsilon)}) - f(\sqrt{t}\, e^{i(\pi/2-\varepsilon)})}{2i} = \sum_{k=0}^{\infty} a_k t^{(k-\alpha)/2} \sin\left[\left(\frac{\pi}{2} - \varepsilon\right)(\alpha - k)\right]$$

$$\ll \sum_{k=0}^{\infty} |a_k| t^{(k-\alpha)/2}, \qquad 0 < t < \rho^2.$$

We may let $\varepsilon \to 0+$ term by term since the dominant series is independent of ε. We obtain

$$\phi(\sqrt{t}) = \sum_{k=0}^{\infty} a_k t^{(k-\alpha)/2} \sin \frac{\pi}{2}(\alpha - k), \qquad 0 < t < \rho^2,$$

$$\phi(t) = \sum_{k=0}^{\infty} a_k t^{k-\alpha} \sin \frac{\pi}{2}(\alpha - k), \qquad 0 < t < \rho.$$

Thus the conclusion of Theorem 6.1 is actually valid throughout the larger interval, as we stated above.

7. Relation to Potential Theory

In §2 we observed that the transform (2.1) under discussion is related to the Poisson integral representation of a function which is harmonic in a half-plane. We elucidate here. It is known, L. V. Ahlfors [1966, p. 169], that a function $u(x, y)$ which is harmonic and bounded for $y > 0$ and continuous for $y \geq 0$ has the integral representation.

$$(7.1) \qquad u(x, y) = \frac{1}{\pi} \int_{-\infty}^{\infty} \frac{y u(t,0)\, dt}{(x - t)^2 + y^2}, \qquad y > 0.$$

If in addition $u(x, y)$ is an even function of x for each $y \geq 0$ then

$$u(0, y) = \frac{1}{\pi} \int_{-\infty}^{\infty} \frac{y u(t, 0)\, dt}{t^2 + y^2} = \frac{2y}{\pi} \int_{0}^{\infty} \frac{t}{t^2 + y^2} \frac{u(t, 0)}{t}\, dt.$$

That is, $u(0, y)/y$ is the potential transform of $u(t, 0)/t$, as we have defined this term in §2. Then by Theorem 5.1

$$(7.2) \qquad L_x\left[\frac{u(0, y)}{y}\right] = \frac{u(x, 0)}{x}.$$

Thus these harmonic functions have the curious property that their values at points of the x axis are obtained, by use of the operator L_x, from their values at corresponding points of the y axis. We state our result as follows.

Theorem 7.

1. $u(x, y) \in C^2.\mathrm{B}, \qquad y \geqq 0$;

2. $\dfrac{\partial^2 u}{\partial x^2} + \dfrac{\partial^2 u}{\partial y^2} = 0, \qquad y > 0$;

3. $u(-x, y) = u(x, y), \qquad y \geqq 0$

$\Rightarrow \qquad L_x\left[\dfrac{u(0, y)}{y}\right] = \dfrac{u(x, 0)}{x}, \qquad x > 0.$

Hypothesis 3 can be dispensed with, but then the conclusion has a more involved statement. (See Exercise 11 of this chapter.)

We illustrate the theorem by the example

$$u(x, y) = e^{-y} \cos x, \qquad L_x\left[\frac{e^{-y}}{y}\right] = \frac{\cos x}{x}.$$

The conclusion of the theorem for this example was also verified by the series algorithm of §4.

8. Relation to the Sine Transform

The sine transform is defined by the equation

$$(8.1) \qquad f(x) = \frac{2}{\pi} \int_0^\infty \sin xt \, \phi(t) \, dt.$$

This reduces to a convolution transform by exponential change of variable:

$$e^x f(e^x) = \frac{2}{\pi} \int_{-\infty}^{\infty} \sin e^{x-t} \, e^{x-t} \phi(e^{-t}) \, dt.$$

Its inversion function $E(s)$ is given by the equation

$$\frac{1}{E(s)} = \frac{2}{\pi} \int_{-\infty}^{\infty} e^{-st} e^t \sin e^t \, dt = \frac{2}{\pi} \int_{0}^{\infty} x^{-s} \sin x \, dx$$

$$= \frac{1}{\Gamma(s) \sin \dfrac{\pi s}{2}} = \frac{2}{\pi} \Gamma(1 - s) \cos \frac{\pi s}{2}.$$

See Erdélyi [1954; 317], formula (1). We thus have the symbolic inversion formula

$$(8.2) \qquad E(D)e^x f(e^x) = \sin \frac{\pi D}{2} \, [\Gamma(D)e^x f(e^x)] = \phi(e^{-x}).$$

But the operator $\Gamma(D)$ is related to the direct Laplace transform and $\sin(\pi D/2)$ to the inverse potential transform. Without belaboring these operational considerations further let us proceed at once to their realization, thus inverting (8.1) in two steps; a Laplace transform of the function $f(x)$ followed by the operator L_x of Definition 5.

Theorem 8.

$$1. \quad \phi(t) \in \text{L.C}, \qquad (0, \infty);$$

$$2. \quad f(x) = \frac{2}{\pi} \int_{0}^{\infty} \phi(t) \sin xt \, dt;$$

$$3. \quad R(x) = \int_{0}^{\infty} e^{-xt} f(t) \, dt$$

$$(8.3) \quad \Rightarrow \qquad L_x[R] = \phi(x), \qquad 0 < x < \infty.$$

It is clear, by virtue of hypothesis 1, that $f(x) \in$ B.C. Hence $R(x)$ is well defined for $x > 0$, so that the first step of the inversion is feasible. By Fubini's theorem we have

$$R(x) = \frac{2}{\pi} \int_0^\infty e^{-xt} \, dt \int_0^\infty \sin ty \, \phi(y) \, dy$$

$$= \frac{2}{\pi} \int_0^\infty \phi(y) \, dy \int_0^\infty e^{-xt} \sin ty \, dt = \frac{2}{\pi} \int_0^\infty \frac{y\phi(y)}{x^2 + y^2} \, dy.$$

Thus $R(x)$ is a potential transform, as suggested by the symbolic equation (8.2). Now Theorem 5.1 is directly applicable and equation (8.3) results.

It should be observed that the classical inversion of (8.1),

$$(8.4) \qquad\qquad \phi(x) = \int_0^\infty \sin xt f(t) \, dt,$$

is not applicable under the hypotheses 1. Compare E. C. Titchmarsh [1937, p.17]. Some stronger local condition on $\phi(t)$ such as bounded variation would be needed. Moreover, it would not be difficult to show that our hypothesis of continuity could be omitted completely, where-upon (8.3) would hold almost everywhere.

As an example consider the equation (Exercise 2 this chapter)

$$(8.5) \qquad\qquad xe^{-x} = \frac{2}{\pi} \int_0^\infty \sin xt \, \frac{2t \, dt}{(t^2 + 1)^2} \, .$$

The function $\phi(t) = 2t/(t^2 + 1)^2$ satisfies hypothesis 1 of Theorem 8, and

$$R(x) = \int_0^\infty e^{-xt} te^{-t} \, dt = \frac{1}{(x + 1)^2}, \qquad x > 0.$$

This function is a potential transform which we may invert by the series method, as follows:

$$\frac{1}{(x + 1)^2} = 1 - 2x + 3x^2 - 4x^3 + \cdots,$$

$$\phi(x) = 2x - 4x^3 + 6x^4 - \cdots = -\frac{d}{dx}(x^2 + 1)^{-1} = \frac{2x}{(x^2 + 1)^2} \, .$$

We have thus inverted (8.5) by the two steps of Theorem 8. Of course for this simple case the classical inversion (8.4) is also valid:

$$\frac{2x}{(x^2 + 1)^2} = \int_0^\infty te^{-t} \sin xt \, dt,$$

since $xe^{-x} \in L$ on $(0, \infty)$. See Erdélyi [1954, p.72], formula (3).

9. The Laplace Transform

Let us show briefly how the method of series used for the potential transform can also be applied to the Laplace transform. If we choose

$$K(x) = \frac{e^{-1/x}}{x}$$

in the integral (1.1) we obtain

$$(9.1) \qquad f(x) = \frac{1}{x} \int_0^\infty e^{-t/x} \phi(t) \, dt.$$

This is the Laplace transform, in which the variable x of the generating function has been replaced by its reciprocal.

For the inversion function we have

$$\frac{1}{E(s)} = \int_0^\infty e^{-1/x} x^{s-2} \, dx = \Gamma(1-s).$$

Let us use the Euler product expansion for the gamma-function,

$$(9.2) \qquad \frac{1}{\Gamma(1-s)} = \lim_{n \to \infty} n^s \prod_{k=1}^n \left(1 - \frac{s}{k}\right).$$

See E. T. Whittaker and G. N. Watson [1943, p.237], for example. As in §5 we are thus led to make the following definition of our inversion operator. As before

$$(9.3) \qquad \theta[f(x)] = - xf'(x).$$

Definition 9.

$$L_{n,x}[f] = n^\theta \prod_{k=1}^n \left(1 - \frac{\theta}{k}\right) f(x),$$

$$L_x[f] = \lim_{n \to \infty} L_{n,x}[f].$$

Except for the factor n^θ, $L_{n,x}$ would be an Euler differential operator of order n. We could write (9.3) as

$$\theta[f(x)] = Df(e^{-y}) \Big|_{y=-\log x},$$

where D stands for differentiation with respect to y. Accordingly n^θ should be interpreted as

$$n^\theta[f(x)] = e^{(\log n)\theta}[f] = e^{(\log n)D}f(e^{-y})\Big|_{y=-\log x}.$$

Recalling our interpretation in §3, Chapter 7, of e^{aD} as a translation we have

$$n^\theta[f(x)] = f(e^{-y-\log n})\Big|_{y=-\log x} = f\left(\frac{x}{n}\right).$$

This is the meaning of n^θ to be adopted in Definition 9. For example,

$$e^{n\theta}x^\alpha = \left(\frac{x}{n}\right)^\alpha$$

for any real number α and any positive number n.

With this understanding we obtain

$$L_{n,x}[x^\alpha] = S_n(-\alpha)x^\alpha,$$

where

(9.4)
$$S_n(-\alpha) = n^{-\alpha}\prod_{k=1}^{n}\left(1+\frac{\alpha}{k}\right).$$

By (9.2)

(9.5)
$$\lim_{n\to\infty} S_n(-\alpha) = \frac{1}{\Gamma(1+\alpha)}.$$

Let us now apply the operation $L_{n,x}$ to the integral (9.1). One shows by induction that

(9.6)
$$L_{n,x}\left[\frac{e^{-t/x}}{x}\right] = \left(\frac{n}{x}\right)^{n+1}e^{-nt/x}t^n.$$

Thus

$$L_{n,x}[f] = \left(\frac{n}{x}\right)^{n+1}\int_0^\infty [e^{-t/x}t]^n\phi(t)\,dt,$$

and the Laplace asymptotic method is clearly applicable. We are led to the following result.

Theorem 9.

$$1. \quad \phi(t) \in \text{B.C.}, \qquad 0 < t < \infty;$$

$$2. \quad f(x) = \frac{1}{x} \int_0^\infty e^{-t/x} \phi(t) \, dt$$

$$\Rightarrow \qquad L_x[f] = \phi(x), \qquad 0 < x < \infty.$$

For, a change of variable gives

$$L_{n,x}[f] = \frac{n^{n+1}}{n!} \int_0^\infty (e^{-x} x)^n \phi(xy) \, dy.$$

But the limit which this integral approaches as $n \to \infty$ was already computed in §3 of Chapter 6 and found to be $\phi(x)$, $0 < x < \infty$. This completes the proof. As an example consider the pair

$$f(x) = \frac{1}{x+1}, \qquad \phi(t) = e^{-t}.$$

for which $L_{n,x}[f]$ can be computed explicitly:

$$L_{n,x}\left[\frac{1}{x+1}\right] = \left(1 + \frac{x}{n}\right)^{-n-1}.$$

This clearly tends to e^{-x}, as predicted.

10. Series Inversion of the Laplace Transform

As a preliminary to our proof of the following theorem let us show that the function (9.4),

$$S_n(-k) = n^{-k} \prod_{j=1}^n \left(1 + \frac{k}{j}\right),$$

is a decreasing function of n for each positive integer k.
That is,

$$(10.1) \qquad S_{n+1}(-k) \leqq S_n(-k), \qquad n, k = 1, 2, \ldots,$$

$$(n+1)^{-k}\left(1 + \frac{k}{n+1}\right) \leqq n^{-k},$$

$$\left(1 + \frac{k}{n+1}\right) \leqq \left(1 + \frac{1}{n}\right)^k.$$

This last inequality becomes evident if we apply the binomial expansion to the right-hand side. The left-hand term is less than the sum of the first two terms of the expansion into $k + 1$ *positive* terms.

We are now in a position to prove the following result.

Theorem 10.1.

$$1. \quad \phi(t) \in \text{B.C}, \qquad 0 < t < \infty;$$

$$2. \quad f(x) = \frac{1}{x} \int_0^\infty e^{-t/x} \phi(t) \, dt;$$

$$3. \quad f(x) = \sum_0^\infty a_k x^k, \qquad 0 < x < \rho$$

$$(10.2) \quad \Rightarrow \qquad \phi(x) = \sum_{k=0}^\infty a_k \frac{x^k}{k!}, \qquad 0 < x < \infty.$$

Since power series may be differentiated term by term we have at once from hypothesis 3 that

$$(10.3) \quad L_{n,x}[f] = \sum_{k=0}^\infty a_k S_n(-k) x^k, \qquad 0 < x < n\rho.$$

Note the enlarged interval of convergence for this series, due to the factor n^{-k} in $S_n(-k)$.

Let x_1 be an arbitrary positive number and choose $n_1 > x_1/\rho$. The monotonic character of the function $S_n(-k)$ shows that

$$\sum_{k=1}^\infty a_k S_n(-k) x_1{}^k \ll \sum_{k=1}^\infty |a_k| S_{n_1}(-k) x_1{}^k, \qquad n \geq n_1.$$

The dominant series converges since $x_1 < n_1\rho$, and consequently the series on the left converges uniformly over the integers n greater than n_1. This fact permits us to let n become infinite term by term in (10.2) to obtain, by (9.5), that

$$L_{x_1}[f] = \sum_{k=0}^\infty a_k \frac{x_1{}^k}{k!}.$$

The sum of this series is $\phi(x_1)$ by Theorem 9. Since x_1 was arbitrary the proof is complete.

As an example consider the pair $\phi(t) = \sin t$,

$$f(x) = \frac{x}{x^2 + 1} = x - x^3 + x^5 - \cdots, \qquad 0 < x < 1.$$

The transformed series (10.2) becomes

$$x - \frac{x^3}{3!} + \frac{x^5}{5!} - \cdots,$$

which has its sum equal to sin x, as predicted.

It should be pointed out that for the Laplace kernel an alternate method of proof is available. Term by term integration of series (10.2) may be made to yield the desired result. (See Exercise 15 of this chapter.) But note that such a method was not available for the proof of Theorem 6.1 since the potential transform of t^n is not defined for any integer n.

We turn next to the case in which the generating function can be expanded in a series of negative powers. For its proof we shall need the following preliminary result.

Lemma 10. For $\alpha \geq 1$ and $n = 1, 2, 3, \ldots$

$$(10.4) \qquad\qquad |S_n(\alpha)| < e^{\pi\alpha}\Gamma(\alpha).$$

Although (10.1) was proved for integers $k \geq 1$ it is equally true for arbitrary $\alpha \geq 1$. (See Exercise 18 of this chapter.) Hence by (9.5)

$$\frac{1}{\Gamma(\alpha + 1)} \leq S_n(-\alpha), \qquad \alpha \geq 1, \quad n = 1, 2, 3, \ldots.$$

But

$$S_n(\alpha) = n^\alpha \prod_{k=1}^{n} \left(1 - \frac{\alpha}{k}\right) = \frac{\displaystyle\prod_{k=1}^{n} \left(1 - \frac{\alpha^2}{k^2}\right)}{S_n(-\alpha)},$$

$$|S_n(\alpha)| \leq \Gamma(1 + \alpha) \prod_{k=1}^{n} \left(1 + \frac{\alpha^2}{k^2}\right) < \Gamma(1 + \alpha) \frac{\sinh \pi\alpha}{\pi\alpha}.$$

From the functional equation for the gamma-function and from the definition of sinh α inequality (10.4) now follows at once.

We can now prove the following result.

Theorem 10.2.

$$1. \quad \phi(t) \in \text{B.C}, \qquad 0 < t < \infty;$$

$$2. \quad f(x) = \frac{1}{x} \int_0^\infty e^{-t/x} \phi(t)\, dt;$$

$$3. \quad f(x) = \sum_{k=0}^{\infty} a_k \, x^{-k-\alpha}, \qquad \text{some } \alpha;$$

$$4. \qquad a_k = O\left(\frac{r^k}{k!}\right), \qquad k \to \infty, \quad \text{some } r > 0$$

$$\Rightarrow \qquad \phi(x) = \sum_{k=0}^{\infty} \frac{a_k x^{-k-\alpha}}{\Gamma(1-k-\alpha)}, \qquad re^{\pi} < x < \infty.$$

Under hypothesis 4 the series expansion for $f(x)$ clearly converges for all $x > 0$. Hence

$$(10.5) \qquad L_{n,x}[f] = \sum_{k=0}^{\infty} a_k \, S_n(\alpha + k) x^{-\alpha-k}, \qquad 0 < x < \infty.$$

By Lemma 10 and hypothesis 4

$$\sum_{\alpha+k \geq 1} a_k \, S_n(\alpha + k) x^{-\alpha-k} \ll M \sum_{\alpha+k \geq 1} \frac{\Gamma(\alpha+k)}{k!} r^k e^{\pi(\alpha+k)} x^{-\alpha-k}.$$

The test ratio for the dominant series is

$$\lim_{k \to \infty} \frac{\Gamma(\alpha+k+1)re^{\pi}}{\Gamma(\alpha+k)(k+1)x} = \frac{re^{\pi}}{x}.$$

Accordingly it converges for $x > re^{\pi}$. Since the dominant series is independent of n the series (10.5) converges uniformly in n for large n. Allowing n to become infinite therein we obtain the desired result by use of equation (9.5).

As an illustration take the determining function $\phi(t)$ equal to unity for $t > 1$ and to zero for $t < 1$. The corresponding generating function is

$$f(x) = \sum_{k=0}^{\infty} \frac{(-1)^k}{k!} x^{-k}.$$

By Theorem 10.2

$$L_x[f] = \phi(x) = \sum_{k=0}^{\infty} \frac{(-1)^k x^{-k}}{\Gamma(1-k)}, \qquad e^{\pi} < x < \infty.$$

But all terms of this series, except the first, vanish and the sum is unity, as expected. Note that the inversion of the theorem is valid only in a neighborhood of infinity, as this example shows.

For a less trivial example take

$$\phi(t) = [\pi(t-1)]^{-1/2}, \quad t > 1$$
$$= 0, \quad t < 1,$$
$$f(x) = \frac{e^{-1/x}}{\sqrt{x}} = \sum_{k=0}^{\infty} \frac{(-1)^k}{k!} x^{-k-1/2}.$$

By Theorem 10.2

$$(10.6) \qquad \phi(x) = \sum_{k=0}^{\infty} \frac{(-1)^k}{k!} \frac{x^{-k-1/2}}{\Gamma\left(\frac{1}{2}-k\right)}.$$

But

$$\frac{\pi}{\Gamma\left(\frac{1}{2}-k\right)} = (\cos \pi k)\Gamma\left(\frac{1}{2}+k\right) = (-1)^k \frac{(2k)!}{k!} \frac{\sqrt{\pi}}{4^k}.$$

Hence (10.6) becomes

$$\phi(x) = \frac{1}{\sqrt{\pi x}} \sum_{k=0}^{\infty} \binom{2k}{k} \frac{1}{(4x)^k} = \frac{1}{\sqrt{\pi}\sqrt{x-1}}.$$

The theorem is seen to yield the correct result, at least for x sufficiently large.

11. Summary

In this chapter we have treated the inversion of the transform

$$f(x) = \int_0^{\infty} K\left(\frac{x}{t}\right) \frac{\phi(t)}{t}\, dt$$

for two special cases of the kernel K. We have shown that when the generating function has a power series expansion,

$$(11.1) \qquad (x)f = \sum_{k=0}^{\infty} a_k x^k,$$

for example, then the determining function can be obtained therefrom by the insertion of a set of multipliers into the series (11.1):

$$\phi(x) = \sum_{k=0}^{\infty} a_k E(-k)x^k.$$

We have chosen the kernels K in the two examples in such a way as to make the multipliers of a specially simple and interesting nature. However, the method used is general and can be applied to a large variety of kernels. See D. V. Widder [1968].

EXERCISES

1. Prove that the potential transform of $\sin at$ is e^{-ax}.

2. Differentiate the transform of the previous exercise with respect to x and set $x = 1$ to obtain equation (8.5).

3. If $f(x)$ is the transform of $\phi(t)$ by equation (1.1) prove that $xf'(x)$ is the transform of $t\phi'(t)$ under suitable conditions, which you are to obtain.
    [Hint: Integrate by parts.]

4. Use the previous exercise to obtain the potential transform of $(t^3 - t)/(t^2 + 1)^2$.

    $$\text{Ans.} \quad \frac{x}{(x^2 + 1)^2}.$$

5. Prove that the potential transform of $2t/(t^2 + 1)^2$ is $1/(x + 1)^2$.

6. If $z = x + iy$ show that $u(x, y) = \operatorname{Im} z/(1 - iz)$ satisfies the conditions of Theorem 7. Use equation (7.2) to establish the transform A of §3.

7. Obtain Formula B of §3 from

    $$u(x, y) = \operatorname{Im} ze^{iz}.$$

8. Obtain Formula C of §3 from

    $$u(x, y) = \operatorname{Re} e^{iz}.$$

9.   Show that the function

$$u(x, y) = 1 - \frac{1}{\pi} \cot^{-1} \frac{x-1}{y} + \frac{1}{\pi} \cot^{-1} \frac{x+1}{y}$$

is harmonic and bounded for $y > 0$. It has a Poisson representation (7.1). Apply equation (7.2) to obtain Formula D of §3.

10.   Obtain the transform pair of Exercise 5 by use of the function

$$u(x, y) = \operatorname{Im} \frac{z}{(1 - iz)^2}.$$

12.   Show that if condition 3 of Theorem 7 is omitted that equation (7.2) should be replaced by

$$L_x\left[\frac{u(0, y)}{y}\right] = \frac{[u(x, 0) + u(-x, 0)]}{2x}.$$

12.   Show by any method that $1/(x^2 + x)$ is the potential transform of $1/(t^3 + t)$.

13.   Prove Theorem 6.2.

14.   Prove Formula (9.6).

15.   Prove Theorem 10.1 by use of Theorem 12 of Chapter 5.

16.   Generalize Theorem 10.1 to include the case when $f(x)$ has an expansion

$$f(x) = \sum_{k=0}^{\infty} a_k x^{k+\alpha}, \qquad \text{some } \alpha.$$

17.   Illustrate the previous exercise by use of the pair

$$f(x) = \frac{e^{-x}}{\sqrt{x}}, \qquad \phi(t) = (\pi t)^{-1/2} \cosh (4t)^{1/2}.$$

See Erdélyi [1954, p.245], formula (32).

18.   Prove inequality (10.1) when k is an arbitrary real number ≥ 1. [Hint: If $f(\alpha) = (1 + 1/n)^\alpha/(1 + \alpha/n)$, then $f'(\alpha) > 0$ when $\alpha > 1$.]

19.   Show that in Theorem 10.2 hypothesis 4 may be replaced by "$f(1/x)$ is entire of growth $\{1, r\}$, some $r > 0$". See Definition 12 of Chapter 5.

Bibliography

Abel, N. H.
 1826. Auflösing einer mechanischen Aufgaben. *Crelle Journal*, Vol. **1**, pp. 153–157.

Bernstein, S.
 1928. Sur les fonctions absolument monotones. *Acta Mathematica*, Vol. **51**, pp. 1–66.

Boas, R. P.,
 1954. "Entire functions." Academic Press, New York.

Bohr, H.
 1911. Beweis der Existenz Dirichletscher Reihen, die Nullstellen mit beliebig grosser Abszisse besitzen. *Rendiconti del Circolo Matematico di Palermo*, Vol. **31**, pp. 235–243.

Cahen, E.
 1894. Sur la fonction $\zeta(s)$ de Riemann et sur les fonctions analogues. *Annales Scientifiques de l'École Normale Supérieure*, Vol. **11**, pp. 75–164.

Carleman, T.
 1926. "Les fonctions quasi analytiques." Gauthier-Villars, Paris.

Comrie, L. J.
 1936. See Titchmarsh, E. C., 1936.

Erdélyi, A.
 1954. "Table of Integral Transforms" (Bateman Manuscript Project), Vol. **1**. McGraw-Hill, New York.

Gram, J. P.
  1903. Notes sur les zéros de la fonction $\zeta(s)$ de Riemann. *Acta Mathematica*, Vol. 27, pp. 289–304.

Hamburger, H.
  1921. Über die Riemannsche Funktionalgleichung der ζ-Funktion. *Mathematische Zeitschrift*, Vol. **10**, pp. 240–254.

Hardy, G. H.
  1910. Theorems relating to the summability and convergence of slowly oscillating series. *Proceedings of the London Mathematical Society* (2), Vol. **8**, pp. 301–320.
  1914. Sur les zéros de la fonction $\zeta(s)$ de Riemann. *Comptes Rendus des Séances de L'Académie des Sciences*, Vol. **158**, pp. 1012–1014.
  1929. An introduction to the theory of numbers. *Bulletin of the American Mathematical Society*, Vol. **35**, pp. 778–818.

Hardy, G. H. and Riesz, M.
  1915. "The General Theory of Dirichlet's Series." Cambridge University Press, Cambridge. (Cambridge Tract No. 18).

Hausdorff, F.
  1921. Summationsmethoden und Momentfolgen. I. *Mathematische Zeitschrift*, Vol. **9**, pp. 74–109.

Hirschman, I. I. and Widder, D. V.
  1955. "The Convolution Transform." Princeton Univ. Press, Princeton, New Jersey.

Karamata, J.
  1931. Neuer Beweis und Veralgermeinerung der Tauberschen Sätze, welche die Laplacesche und Stieltjessche Transformation betreffen. *Journal für die reine und angewandte Mathematik*, Vol. **164**, pp. 27–39.

Knopp. K.
  1928. "Theory and Application of Infinite Series." Blackie and Son, London.

Laguerre, E.
  1882. Sur les fonctions du genre zéro et du genre un. *Comptes Rendus des Séances de l'Académie des Sciences*, Vol. **98**, pp. 828–831.

Landau, E.
  1905. Über ein Satz von Tschebyschef. *Mathematische Annalen*, Vol. **61**, pp. 527–550.

Lerch, M.
  1903. Sur un point de la théorie des fonctions génératrices d'Abel. *Acta Mathématica*, Vol. **27**, pp. 339–351.

Levinson, N.
  1969. A motivated account of an elementary proof of the prime number theorem. *American Mathematical Monthly*, Vol. **76**, pp. 225–245.

Lindelöf, E.
  1908. Quelques remarques sur la croissance de la fonction $\zeta(s)$. *Bulletin des Sciences Mathématiques*, Vol. **32**, pp. 341–356.

Littlewood, J. E.
   1910. The converse of Abel's theorem on power series. Proceedings of the London Mathematical Society, Vol. 9, pp. 434–438.

Müntz, C.
   1914. Über den Approximationssatz von Weierstrass, pp. 303–312. Schwarz's Festschrift. Springer, Berlin.

Pólya, G.
   1913. Über Annäherung durch Polynome mit lauter reelen Wurlzeln. *Rendiconti del Circolo Matematico di Palermo*, Vol. **36**, pp. 279–295.

Riemann, B.
   1876. Über die Anzahl der Primzahlen unter einer gegebener Grösse. Gesammelte Mathematische Werke, Teubner, Leipzig.

Riesz, M.
   See Hardy, G. H.

Selberg, A.
   1942. On the zeros of Riemann's zeta-function on the critical line. *Archiv for Mathematik og Naturvidenskab*, Vol. **45**, pp. 101–114.

Skewes, S.
   1955. On the difference $\pi(x) - \mathrm{li}(x)$, II. *Proceedings of the London Mathematical Society* (3), Vol. **5**, pp. 48–70.

Stieltjes, T. J.
   1894. Recherches sur les fractions continues. *Annales de la Faculté des Sciences de Toulouse*, Vol. **8**, pp. 1–122.

Tagamlitzki, Y.
   1946. Sur les suites verifiant certaines inegalités. *Comptes Rendus des Séances de l'Académie des Sciences*, Vol. **223**, pp. 940–942.

Tauber, A.
   1897. Ein Satz aus der Theorie der unendlichen Reihen. *Monatshefte für Mathematik und Physik*, Vol. **8**, pp. 273–277.

Titchmarsh, E. C.
   1936. The zeros of the Riemann zeta-function. *Proceedings of the Royal Society of London*, Vol. **157**, pp. 261–263.
   1939. "The Theory of Functions." Oxford Univ. Press, London and New York.

Watson, G. N.
   See Whittaker, E. T.

Whittaker, E. T. and Watson, G. N.
   1943. "Modern Analysis." Macmillan, New York.

Widder. D. V.
   1934. The inversion of the Laplace integral and the related moment problem. *Transactions of the American Mathematical Society*, Vol. **36**, pp. 107–200.
   1938. The Stieltjes transform. *Transactions of the American Mathematical Society*, Vol. **43**, pp. 7–60.
   1940. Functions whose even derivatives have a prescribed sign. *Proceedings of the National Academy of Sciences, U.S.* Vol. **26**, pp. 657–659.

1946.   "The Laplace Transform." Princeton University Press, Princeton, New Jersey.
1961.   "Advanced Calculus." Prentice-Hall, Englewood Cliffs, New Jersey. See also Hirschman, I. I.

Wiener, N.
1933.   "The Fourier Integral." Cambridge University Press, Cambridge.

Index

Pure and Applied Mathematics

A Series of Monographs and Textbooks

Editors

Paul A. Smith and Samuel Eilenberg

Columbia University, New York

1: ARNOLD SOMMERFELD. Partial Differential Equations in Physics. 1949 (Lectures on Theoretical Physics, Volume VI)

2: REINHOLD BAER. Linear Algebra and Projective Geometry. 1952

3: HERBERT BUSEMANN AND PAUL KELLY. Projective Geometry and Projective Metrics. 1953

4: STEFAN BERGMAN AND M. SCHIFFER. Kernel Functions and Elliptic Differential Equations in Mathematical Physics. 1953

5: RALPH PHILIP BOAS, JR. Entire Functions. 1954

6: HERBERT BUSEMANN. The Geometry of Geodesics. 1955

7: CLAUDE CHEVALLEY. Fundamental Concepts of Algebra. 1956

8: SZE-TSEN HU. Homotopy Theory. 1959

9: A. M. OSTROWSKI. Solution of Equations and Systems of Equations. Second Edition. 1966

10: J. DIEUDONNÉ. Treatise on Analysis. Volume I, Foundations of Modern Analysis, enlarged and corrected printing, 1969. Volume II, 1970.

11: S. I. GOLDBERG. Curvature and Homology. 1962.

12: SIGURDUR HELGASON. Differential Geometry and Symmetric Spaces. 1962

13: T. H. HILDEBRANDT. Introduction to the Theory of Integration. 1963.

14: SHREERAM ABHYANKAR. Local Analytic Geometry. 1964

15: RICHARD L. BISHOP AND RICHARD J. CRITTENDEN. Geometry of Manifolds. 1964

16: STEVEN A. GAAL. Point Set Topology. 1964

17: BARRY MITCHELL. Theory of Categories. 1965

18: ANTHONY P. MORSE. A Theory of Sets. 1965

Pure and Applied Mathematics

A Series of Monographs and Textbooks